Geophysical Interpretation using Integral Equations

Geophysical Interpretation using Integral Equations

L. ESKOLA

*Head of the Geophysics Department,
Geological Survey of Finland*

CHAPMAN & HALL
London · New York · Tokyo · Melbourne · Madras

Published by Chapman & Hall, 2–6 Boundary Row, London SE1 8HN

Chapman & Hall, 2–6 Boundary Row, London SE1 8HN

Chapman & Hall, 29 West 35th Street, New York NY10001

Chapman & Hall Japan, Thomson Publishing Japan, Hirakawacho Nemoto Building, 6F, 1–7–11 Hirakawa-cho, Chiyoda-ku, Tokyo 102

Chapman & Hall Australia, Thomas Nelson Australia, 102 Dodds Street, South Melbourne, Victoria 3205

Chapman & Hall India, R. Seshadri, 32 Second Main Road, CIT East, Madras 600 035

First edition 1992

©1992. L. Eskola

Typeset in 10/12 pt Times by Pure Tech Corporation, India
Printed in Great Britain by T. J. Press (Padstow) Ltd., Padstow, Cornwall.

ISBN 0 412 37020 4

Apart from any fair dealing for the purposes of research or private study, or criticism or review, as permitted under the UK Copyright Designs and Patents Act, 1988, this publication may not be reproduced, stored, or transmitted, in any form or by any means, without the prior permission in writing of the publishers, or in the case of reprographic reproduction only in accordance with the terms of the licences issued by the Copyright Licensing Agency in the UK, or in accordance with the terms of licences issued by the appropriate Reproduction Rights Organization outside the UK. Enquiries concerning reproduction outside the terms stated here
should be sent to the publishers at the London address printed on this page.
 The publisher makes no representation, express or implied, with regard to the accuracy of the information contained in this book and cannot accept any legal responsibility or liability for any errors or omissions that may be made.

A catalogue record for this book is available from the British Library

Library of Congress Cataloging-in-Publication data available

Contents

Preface ix

Introduction xi

1 General matters concerning integral equations 1
1.1 Demonstration of an integral equation solution 1
1.2 Classification of integral equations 4
1.3 Numerical solution 6
   1.3.1 The method of moments 6
   1.3.2 The Galerkin method 7
   1.3.3 The point-matching method 7

2 Elements of electrostatics and potential theory 9
2.1 Differential representation of electrical potential 9
2.2 Integral representation of electrical potential 12
2.3 Primary current electrode 15
2.4 Volume charge distribution 16
2.5 Surface charge distribution 17
2.6 Electrical double layer 18

3 Electrical methods 21
3.1 Introduction 21
3.2 Resistivity of rocks 22
3.3 Resistivity method 23
   3.3.1 Introduction 23
   3.3.2 Integral equations for charge density 24
   3.3.3 Integral equations for potential 30
   3.3.4 Integral equation for perfect conductor model 33
   3.3.5 Integral equations for two-dimensional earth models 36
   3.3.6 Integrodifferential equations for thin conductor models 42

3.4	Magnetometric resistivity	48
3.5	*Mise-à-la-masse* method	52
	3.5.1 Integral equations for charge density	52
	3.5.2 Integral equation for potential	55
	3.5.3 Integral equation for perfect conductor model	57
3.6	Surface polarization	59
	3.6.1 Introduction	59
	3.6.2 Solution for perfect conductor models with surface polarization	60
	3.6.3 Scaling of surface polarization models	64
3.7	Induced polarization	66
	3.7.1 Introduction	66
	3.7.2 Information-theoretical relations concerning resistivity dispersion	66
	3.7.3 Modelling of IP anomalies	69
	3.7.4 Applications	70
3.8	Self-potential	71
	3.8.1 Origin of potentials	71
	3.8.2 Integral equations for mineralization potential	73
	3.8.3 Integral equations for electrofiltration potential	78
3.9	Electrical anisotropy	82
	3.9.1 Definition	82
	3.9.2 Physical significance	83
	3.9.3 Integral equation for perfect conductor in an anisotropic environment	87
	3.9.4 Integral equation for an anisotropic body in an anisotropic environment	91
4	**Elements of magnetostatics**	94
4.1	Introduction	94
4.2	Integral representation of magnetic potential	94
4.3	Volume distribution of poles	97
4.4	Surface distribution of poles	99
4.5	Volume distribution of dipoles	100
5	**Magnetic methods**	102
5.1	Magnetic properties of rocks	102
5.2	High-susceptibility models	103
	5.2.1 Three-dimensional models	103
	5.2.2 $2\frac{1}{2}$-dimensional models	106
	5.2.3 Thin sheet model	107
5.3	Demagnetization and low-susceptibility models	110
	5.3.1 Introduction	110
	5.3.2 The demagnetization factor	111

	5.3.3 Low-susceptibility models	113
5.4	Numerical applications	116
5.5	Effect of remanence	120

6 Electromagnetic methods — 123

6.1	Introduction	123
6.2	Boundary-value problems for electromagnetic fields	124
	6.2.1 Electric field	124
	6.2.2 Magnetic field	126
6.3	Green's dyadics for electromagnetic boundary-value problems	127
6.4	Volume integral equations for three-dimensional electromagnetic fields	131
	6.4.1 Electric field	131
	6.4.2 Magnetic field	136
6.5	Volume integral equations for two-dimensional electromagnetic fields	139
	6.5.1 Introduction	139
	6.5.2 The $E_\perp$ field	139
	6.5.3 The $E_\parallel$ field	142
6.6	Surface integral equations for electromagnetic fields	144
	6.6.1 Three-dimensional model	145
	6.6.2 $2\tfrac{1}{2}$-dimensional model	148
6.7	Integral equation solution for electromagnetic fields in a thin conductor model	152

7 Seismic methods — 158

7.1	Introduction	158
7.2	Integral formulas for elastic wave fields in an anisotropic medium	159
7.3	Integral formulas for elastic wave fields in an isotropic medium	163
7.4	Separation of elastic wave fields into a compressional and a rotational mode	168
7.5	Integral formulas for acoustic wave fields in the frequency domain	169
7.6	Integral formulas for acoustic wave fields in the time domain	172
7.7	Applications	174
	7.7.1 Scattering	174
	7.7.2 Migration	175

Appendix A — 177
Green's function for scalar potential in a two-layer half-space

Appendix B 180
Green's function for scalar potential in a half-space with a vertical contact

Appendix C 182
Green's function for scalar potential in an anisotropic half-space

Appendix D 184
Electric Green's dyadic for a half-space below the ground surface

References 186

Index 189

Preface

Along with the general development of numerical methods in pure and applied sciences, the ability to apply integral equations to geophysical modelling has improved considerably within the last thirty years or so. This is due to the successful derivation of integral equations that are applicable to the modelling of complex structures, and efficient numerical algorithms for their solution. A significant stimulus for this development has been the advent of fast digital computers.

The purpose of this book is to give an idea of the principles by which boundary-value problems describing geophysical models can be converted into integral equations. The end results are the integral formulas and integral equations that form the theoretical framework for practical applications. The details of mathematical analysis have been kept to a minimum. Numerical algorithms are discussed only in connection with some illustrative examples involving well-documented numerical modelling results. The reader is assumed to have a background in the fundamental field theories that form the basis for various geophysical methods, such as potential theory, electromagnetic theory, and elastic strain theory. A fairly extensive knowledge of mathematics, especially in vector and tensor calculus, is also assumed.

In Chapter 1 the concept of the integral equation is introduced using a simple direct current boundary-value problem. Definitions for various classes of integral equation, and numerical methods used in solving integral equations are briefly discussed. Chapter 2 gives a summary of the elements of potential theory as applied to electrostatic problems. Chapter 3 is concerned with electrical methods in applied geophysics, involving the resistivity and magnetometric resistivity methods as well as the *mise-à-la-masse*, induced polarization, and self-potential methods. Chapter 4 is concerned with the elements of potential theory applied to magnetostatic problems, the treatment being completely analogous to that presented for electric potential problems in the preceding two sections, while Chapter 5 considers the magnetic methods of applied geophysics. Chapter 6 deals with integral equations for electromagnetic induction problems. Finally, Chapter 7 gives integral formulas for elastic and acoustic wave fields to serve as a mathematical framework for forward and inverse seismic modelling.

The idea of writing this book was originally suggested to me by Professor D. S. Parasnis during the EAEG Meeting in Berlin in 1989. I also wish to express my gratitude to him for the scientific editing of the manuscript. I am grateful to my employer, the Geological Survey of Finland, for allowing me to write this book as part of my normal work. I also thank my colleagues and friends who helped in many ways. Of these, Dr Heikki Soininen, Mr Matti Oksama, Mr Pekka Heikkinen and Mr Hannu Hongisto were kind enough to make critical comments on the manuscript, Mrs Sisko Sulkanen and Mr Markku Eskola drew the illustrations, and Dr Peter Ward nursed my English.

Lauri Eskola
Espoo, Finland

Introduction

Geophysical model calculations are important in evaluating field techniques and interpreting geophysical anomalies. Integral equations, which form the subject of this book, are well suited to the modelling of anomalies obtained by electric, magnetic, electromagnetic and seismic methods.

The spatial behaviour of static electric and magnetic fields is described by Poisson's equation or by a particular form of it known as Laplace's equation. Solutions to particular problems are specified by boundary conditions which in turn depend upon the physical and geometrical structure of a given model. In addition to the static fields which they describe exactly, Poisson's and Laplace's equations also describe, with a high degree of accuracy, certain types of time-varying field. An example of this is the induced polarization method, which utilizes such low frequencies of source current that the effect of electromagnetic induction is negligible. The behaviour of the dynamic electromagnetic and elastic fields in space and time is described by electromagnetic and elastic wave equations and relevant boundary conditions.

All solutions of the boundary-value problems described above fall into one of two classes, analytic and numerical. An analytic solution is obtained in the form of an algebraic equation in which values of the parameters defining the field can be substituted. Analytic methods have the advantage that a general solution can be obtained, from which it is in principle possible to gain an overall picture of the effect of various parameters. A numerical solution takes the form of a set of numerical values of the function, and the effect of different parameters must be calculated separately for each set of values. Hence, an overall picture can often be achieved, but at the expense of a great amount of computation.

Analytical solutions to boundary-value problems are restricted to certain simple geometries of earth models in which the discontinuity surfaces of the medium are coincident with constant coordinate surfaces. For modelling more complex earth structures, effective numerical methods are available. There are two basic approaches to numerical modelling, differential equation methods and integral equation methods. A third approach, the hybrid method, is a combination of the differential and integral equation methods. In the differential equation approach, the entire earth is modelled on a grid, a feature which makes differential equation methods preferable for modelling complex geological structures. Integral equation solutions involve more complex mathematics, but their advantage is that unknown source functions are required only for the anomalous regions. Thus, the integral equation approach is preferable in modelling one or a few small anomalous bodies.

A decisive step towards the practical integral equation modelling of geophysical anomalies was taken in transforming the boundary-value problems defined by the partial differential equations and boundary conditions into integral formulas. A basic integral formula for the electromagnetic field was developed by Stratton and Chu in 1939, and Green's whole-space dyadic was given by Levine and Schwinger in 1950. Integral formulae for acoustic waves were derived for the frequency domain by Helmholtz as early as 1860, and for the time domain by Kirchhoff in 1883. Green's dyadic for an elastic wave field was constructed by Morse and Feshbach in 1953. The very intensive development of integral equations and their numerical solutions into effective tools for geophysical interpretation has taken place within the last thirty years in close connection with the development of fast digital computers. A major objective of present-day research is the further development of the flexibility and efficiency of the modelling algorithms for three- and two-dimensional structures excited by three-dimensional sources.

1

General matters concerning integral equations

Physical theories, such as electromagnetic theory and elastic wave theory, are primarily represented by sets of partial differential equations. For any particular problem, such as the apparent resistivity of an anomalous body located in a half-space, the set of differential equations reduces to a boundary-value problem defined by a partial differential equation and the boundary conditions characteristic of the model under consideration. Such problems can be solved using many alternative methods, but in this book the boundary-value problems are converted into equivalent integral equations which can then be solved by standard numerical methods. To illustrate with the aid of a simple example how a boundary-value problem is transformed into an integral equation we shall, in Section 1.1, solve the potential problem for a line current electrode of finite length, embedded in a homogeneous half-space.

1.1 DEMONSTRATION OF AN INTEGRAL EQUATION SOLUTION

Consider a line electrode L of length l embedded in a half-space of homogeneous resistivity ρ (Fig. 1.1). When an electric current of strength I is fed into the earth through electrode L, a line charge distribution is generated on L and a surface charge distribution on the ground surface A. The charge on the ground surface is the image charge due to the electrostatic interaction between the electrode and the ground surface. The potential U due to the current emitted by L satisfies Poisson's equation

$$\nabla^2 U = -\sigma/\varepsilon_0 \qquad (1.1.1)$$

on L. The source term in equation (1.1.1) is the line charge density σ along electrode L generated by the current spreading from L into the environment. Outside the electrode, U satisfies Laplace's equation

2   *Matters concerning integral equations*

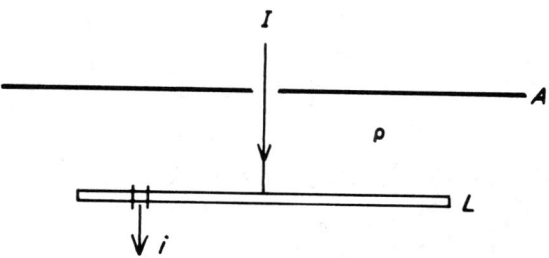

Figure 1.1 Finite line current electrode in a homogeneous half-space.

$$\nabla^2 U = 0 \tag{1.1.2}$$

ε_0 in equation (1.1.1) is a universal constant [Fm^{-1}] required for dimensional consistency in using SI units. Boundary conditions require that the normal component of current density is zero on ground surface A (since the current does not penetrate the earth–air interface), and that $RU(\mathbf{R})$ is bounded as R approaches infinity. Differential equation (1.1.1) and the boundary conditions now define the boundary value problem for potential U.

Next we shall transform the boundary-value problem under consideration into an integral equation with the aid of Green's function technique. Owing to the linearity of the problem, we attempt to express the potential U in the form of the integral

$$U(\mathbf{R}) = \frac{1}{\varepsilon_0} \int_L G(\mathbf{R}, \mathbf{R}_0) \, \sigma(\mathbf{R}_0) \mathrm{d}L_0 \tag{1.1.3}$$

where the calculation point $\mathbf{R}$ is outside the electrode L and source point $\mathbf{R}_0$ is on L. Equation (1.1.3) represents Coulomb's law generalized for a distributed charge. G is Green's function for potential U defined by the differential equation (1.1.1) and any imposed boundary conditions. Since Green's function plays a role parallel to that of the potential, it is natural to require that G satisfy the same basic relations as the potential. Function G is therefore made to satisfy Poisson's equation

$$\nabla^2 G = -\delta(\mathbf{R} - \mathbf{R}_0) \tag{1.1.4}$$

where δ is Dirac's delta function, and boundary conditions similar to those for the potential. Thus, for the present problem, the normal derivative of Green's function (which is proportional to the normal component of current density) must be zero on surface A. To satisfy the regularity condition at infinity, it is required that RG is bounded as R approaches infinity. With these conditions, G is obtained as:

$$G = \frac{1}{4\pi}\left(\frac{1}{|\mathbf{R} - \mathbf{R}_0|} + \frac{1}{|\mathbf{R} - \mathbf{R}_0'|}\right) \tag{1.1.5}$$

where $\mathbf{R}_0'$ is the image of source point $\mathbf{R}_0$ in the ground surface A. The first term of G in equation (1.1.5) corresponds to the potential of the charge distribution $\sigma(\mathbf{R}_0)$ on L, and the second term to the potential of the image charge distributed on ground surface A. It can be seen from equation (1.1.5) that the potential due to a charge distributed on ground surface A is equal on A itself to the potential of the electrical image of electrode L in A.

We have now represented the potential by a convolution of Green's function G and an unknown line charge density σ on L. In order more closely to emphasize the properties of electrode L as a current-emitting source, the rest of the problem is considered in the current regime. First, the unknown function $\sigma(\mathbf{R}_0)$ is expressed in terms of the unknown distribution of line current density $i(\mathbf{R}_0)$ emitted by the electrode. This is made by substituting

$$\sigma(\mathbf{R}_0) = \varepsilon_0 \, \rho \, i(\mathbf{R}_0) \tag{1.1.6}$$

into equation (1.1.3), where ρ is the resistivity of the half-space. Relation (1.1.6) is obtained by using Ohm's law and Gauss's law for the electric field in the environment of L. Derivation of equation (1.1.6) is considered in detail in Section 2.3. The potential U generated by the electrode L can now be written in the form

$$U(\mathbf{R}) = \rho \int_L G(\mathbf{R}, \mathbf{R}_0) \, i(\mathbf{R}_0) \, dL_0 \tag{1.1.7}$$

In an electrical survey with line current electrodes, the current is commonly assumed to be emitted from the electrode with a homogeneous line density $i = I/l$. Using this assumption, the source density (constant) is immediately known and the problem thus solved. In reality, however, the source density is far from constant. A better solution is obtainable by assuming that the potential is constant on electrode L. The quality of the solution obtained with this assumption depends upon the product of the longitudinal conductance of electrode L and the resistivity ρ of the environment. We use now the equipotential condition on L for solving current density $i(\mathbf{R}_0)$ emitted from the electrode. For this purpose, equation (1.1.7) is taken to be satisfied on the electrode in such a manner that the potential obtained is constant on L:

$$U_{\text{const}}(L) = \rho \int_L G(\mathbf{R}, \mathbf{R}_0) \, i(\mathbf{R}_0) \, dL_0 \tag{1.1.8}$$

where $\mathbf{R}$ and $\mathbf{R}_0$ are both on L. However, the potential approaches infinity as the calculation point $\mathbf{R}$ approaches L. We avoid this singularity problem by simply taking equation (1.1.8) as satisfied, in place of electrode L itself, on a line adjacent to L.

Since U_{const} on the left-hand side of equation (1.1.8) is unknown, an additional equation is needed for $i(\mathbf{R}_0)$ in order to make the number of equations equal to the number of the unknowns to be solved. As the supplementary equation we use the conservation relation for current i:

4 *Matters concerning integral equations*

$$\int_L i(\mathbf{R}_0)\,dL_0 = I \qquad (1.1.9)$$

The set of equations (1.1.8) and (1.1.9) is solved to obtain the distribution of $i(\mathbf{R}_0)$ on L. This is made by transforming the equations into a set of linear algebraic equations by discretization. The set is finally solved numerically by elimination or iteration. As soon as $i(\mathbf{R}_0)$ is known, the potential U can be calculated at an arbitrary point of the half-space by equation (1.1.7).

It can be seen from the preceding illustration that the determination of Green's function is a fundamental problem in the integral equation solution for a boundary-value problem. Green's function gives the response due to a unit spike in the source which, once it is known, enables the response to an arbitrary source distribution to be obtained by direct integration. The crucial point is that the analytic solution of Green's function in multi-dimensional problems of geophysics is usually considerably simpler than that of the original boundary-value problem because, in contrast to the original problem, the boundary conditions associated with the anomalous region are not involved in the analytical solution of Green's function.

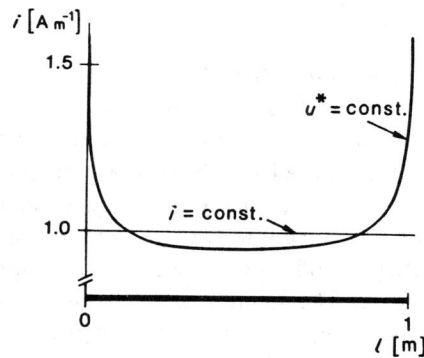

Figure 1.2 Current density due to line electrode located on ground surface, with electrode potential and current density assumed constant. After [1].

As an application of the preceding discussion, the current density emitted by a line electrode situated on ground surface is illustrated in Fig. 1.2. The current density obtained by assuming constant potential on the electrode varies considerably along the electrode, which should be taken into consideration when interpreting field data obtained using line current electrodes.

1.2 CLASSIFICATION OF INTEGRAL EQUATIONS

Any equation in which the unknown function to be determined appears under the sign of integration is called an **integral** equation. The fundamental description of integral equations was given by Fredholm and Volterra. Hilbert and

Schmidt contributed to the development of the theory for integral equations which possess symmetric **Green's functions**. The Green's function is commonly referred to in mathematical theory as the **kernel**.

If the function to be determined occurs only linearly, then the equation is linear. If, in a linear integral equation, the limits of the integral are constant, the equation is of Fredholm's type.

Conversion of the boundary-value problems treated in this book generally leads to **Fredholm integral equations of the second kind**:

$$f(\mathbf{R}) - \lambda \int_V G(\mathbf{R}, \mathbf{R}_0) f(\mathbf{R}_0) \, dV_0 = g(\mathbf{R}) \qquad (1.2.1)$$

where g and G are known functions, and f is the unknown function to be determined. The variable $\mathbf{R}$, like $\mathbf{R}_0$, takes on all values in V. G is Green's function of the integral equation. If in equation (1.2.1) $g(\mathbf{R}) = 0$, the equation is homogeneous.

Green's functions or their derivatives may contain some discontinuity. If the Green's function is singular in the region of integration, the equation is said to be singular. Green's functions appearing in this book are so-called **weakly singular kernels** where integration domains are sufficiently smooth. A sophisticated mathematical theory exists for integral equations with weakly singular kernels, which gives a solid basis for the solution of these equations [2, 3].

If the equation can be solved only for certain values of parameter λ, the problem is an **eigenvalue** problem. If λ has a fixed value, and for this value the equation has a unique solution, the problem is **deterministic**. In the problems considered in this book, λ is uniquely determined by the physical properties of the model, and the problems dealt with are deterministic.

If the unknown function f does not appear outside the integral, the equation is of the form:

$$\int_V G(\mathbf{R}, \mathbf{R}_0) f(\mathbf{R}_0) \, dV_0 = g(\mathbf{R}) \qquad (1.2.2)$$

which is known as a **Fredholm integral equation of the first kind**.

The theory of integral equations of the second kind is considerably more advanced than that for those of the first kind, although in practice, integral equations of the second kind are much easier to handle than those of the first kind. Consequently, it is indeed fortunate that most modelling problems in applied geophysics can be reduced to the former rather than to the latter. Important relations concerning the existence of solutions for integral equations of the second kind are provided by the Fredholm theorems; for a detailed representation of these see Petrovsky [4].

For integral equations of the first kind, the smoother the Green's function G, the more difficult it is to solve such an equation numerically. Even when the solution exists, the system may be unstable, which means that a slight perturbation in g may cause an arbitrarily large variation in f. However, the

6 *Matters concerning integral equations*

Green's functions of the integral equations appearing in this book are infinite at the point $\mathbf{R} = \mathbf{R}_0$, which makes it possible to solve the equations of the first kind with sufficient stability.

1.3 NUMERICAL SOLUTION

A sophisticated method for converting integral equations into sets of algebraic equations is the method of moments, of which a detailed treatment is given in Harrington [5]. The set of equations thus obtained can then be solved by standard numerical methods, namely, elimination or iterative techniques.

1.3.1 The method of moments

Consider the inhomogeneous equation

$$Lf = g \qquad (1.3.1)$$

where L is the linear integral operator of the integral equation, g is the known function and f is the unknown function to be solved. For equation (1.2.1), L is:

$$L = I - \lambda \int_V G \, dV_0$$

where I is the identity tensor, or, for scalar equations, unity. For equation (1.2.2),

$$L = \int_V G \, dV_0$$

The function f is represented by a series of functions f_n, $n = 1, 2, \ldots$, as follows:

$$f = \sum_n a_n f_n \qquad (1.3.2)$$

The coefficients a_n are unknown constants and the f_n are called basis functions. For exact solutions, equation (1.3.2) is usually an infinite summation and the f_n form a complete set of basis functions. In practice, finite summations are generally used, which gives an approximate solution of the original equation. Substituting equation (1.3.2) into equation (1.3.1) results in:

$$\sum_n a_n L f_n = g \qquad (1.3.3)$$

Assume that a suitable inner product $< \cdot , \cdot >$ has been determined for the problem. An example of a commonly applied inner product is:

$$<f, g> = \int_V fg \, dV$$

Now a set of test functions $w_m, m = 1, 2, \ldots$, is defined. Taking the inner product of equation (1.3.3) with each w_m gives:

$$\sum_n a_n <w_m, Lf_n> = <w_m, g> \qquad (1.3.4)$$

which is a set of algebraic equations deduced from the original integral equation (1.3.1). For brevity equation (1.3.4) is written in matrix form:

$$J_{mn} a_n = g_m \qquad (1.3.5)$$

where a_n is the column vector of the unknown coefficients a_n. The elements of the coefficient matrix are $J_{mn} = <w_m, Lf_n>$, and the elements of the known vector on the right-hand side of the equation are $g_m = <w_m, g>$. The unknown coefficients a_n can be solved from the set of linear algebraic equations (1.3.5). The solution for function f is finally obtained from equation (1.3.2) by substitution of coefficients a_n.

In applying the method of moments for any particular problem, the proper choice of functions f_n and w_n is important for the facility and accuracy of the solution. The f_n should be linearly independent and such that some superposition (equation (1.3.2)) approximates the unknown function f sufficiently well. The w_n should be linearly independent and chosen such that the inner products $<w_n, g>$ depend on properties of g.

There are certain specific choices of functions f_n, w_n, which are particularly useful in geophysical applications. Descriptions are given below for the Galerkin method and the point-matching method.

1.3.2 The Galerkin method

If the basis functions f_n and the test functions w_n are chosen such that

$$\mathbf{f}_n = \mathbf{w}_n \qquad (1.3.6)$$

for all values of n, the method is called the Galerkin method.

1.3.3 The point-matching method

In geophysical prospecting the main interest is in the anomaly fields outside the source region V. Thus, the solution of integral equations does not necessarily need high accuracy for details of V. For example, the source density at the corner points of a prism model is singular, but it can be properly truncated if care is taken that the source distribution as a whole is approximated with sufficient accuracy. Low-degree approximations can thus usually be used in geophysical modelling.

A significant simplification is achieved in the integrations of the coefficient matrix by using the point-matching method, in which equation (1.3.1) is required to be satisfied only at discrete points of V. The total number of points must be equal to the number of basis functions f_n. Approximation equation

(1.3.5) is thus obtained, which is solved to obtain the coefficients of the basis functions. This procedure means that the test functions are Dirac delta functions. If the basis functions are also Dirac delta functions, a model is obtained where the source is represented by a discrete distribution of point sources. This method is a special case of both the Galerkin method and the point-matching method.

A special case of the point-matching method is the method of subsections, in which the source region V is divided into elements (or subsections). Each of the basis functions f_n exists only over the elements. A common choice for f_n and w_n is either the triangle function or the pulse function for f_n, and the Dirac delta function for w_n. This choice leads to the point-matching method with subsectional bases. Since this method, using pulse functions as basis functions, is particularly important in geophysical applications of integral equations, it is represented here in more detail.

Let region V be divided into N elements v_i, $i = 1, 2, \ldots, N$. The positions of the elements are characterized by points $\mathbf{R}_i$ (for example, the centre of gravity of the elements). The unknown function f in each element is approximated by constant unknown values $f(\mathbf{R}_i) = f_i$. Green's function for the subsectional system is obtained as follows:

$$G_{ij} = \int_{V_j} G(\mathbf{R}_i, \mathbf{R}_0) \, dV_0 \tag{1.3.7}$$

Equation (1.2.1) is now approximated by the set of algebraic equations:

$$f_i - \sum_{j=0}^{N} G_{ij} f_j = g_i \tag{1.3.8}$$

$i, j = 1, 2, \ldots, N$. Equation (1.3.8) can then be solved in relation to f_i by standard numerical methods.

2

Elements of electrostatics and potential theory

In electrical surveys, resistivity variations distort the current pattern in the earth and so cause anomalies in the potential to be measured. For each current system there exists a uniquely equivalent, physically real system of charge distributions. Since the true sources of static electric field are only charges, this system provides a clear physical basis for using the potential theory in the mathematical formulation of electrical models. An important empirical justification for this is the classical Coulomb law for the field generated by a point charge. In this chapter we shall apply the potential theory in considering the effects generated by the most common source distributions of electrical models. The fundamental bases for this approach are Maxwell's equations, which include Coulomb's law in the form of Gauss's law for the electric field.

2.1 DIFFERENTIAL REPRESENTATION OF ELECTRICAL POTENTIAL

The electromagnetic field theory is based on Maxwell's equations. Using the conventional notation for field vectors where the electric field strength is denoted by $\mathbf{E}\,[\text{V m}^{-1}]$, magnetic flux density by $\mathbf{B}\,[\text{V s m}^{-2}]$, electric flux density by $\mathbf{D}\,[\text{C m}^{-2}]$, magnetic field strength by $\mathbf{H}\,[\text{A m}^{-1}]$, free charge density by $\sigma_F\,[\text{C m}^{-3}]$ and current density by $\mathbf{j}\,[\text{A m}^{-2}]$, Maxwell's equations are stated in differential form as in Stratton [6]:

$$\nabla \times \mathbf{E} = -\frac{\partial \mathbf{B}}{\partial t} \qquad (2.1.1)$$

$$\nabla \times \mathbf{H} = \mathbf{j} + \frac{\partial \mathbf{D}}{\partial t} \qquad (2.1.2)$$

10 *Electrostatics and potential theory*

$$\nabla \cdot \mathbf{B} = 0 \tag{2.1.3}$$

$$\nabla \cdot \mathbf{D} = \sigma_f \tag{2.1.4}$$

$$\nabla \cdot \mathbf{j} = -\frac{\partial \sigma_f}{\partial t} \tag{2.1.5}$$

Equations (2.1.1), (2.1.2) and (2.1.5) are termed independent equations, while equations (2.1.3) and (2.1.4) are deducible from equations (2.1.1), (2.1.2) and (2.1.5) and should thus be considered as dependent equations.

The boundary conditions which must be satisfied by the electric and magnetic fields at physical interfaces, can be deduced from Maxwell's equations, the details of which are presented in any standard textbook on electromagnetism. The most important conditions satisfied by the field vectors at the interface between media 1 and 2 are:

$$E_{1t} = E_{2t}$$

$$j_{1n} - j_{2n} = -\partial \sigma_f / \partial t$$

$$D_{1n} - D_{2n} = \sigma_f$$

$$H_{1t} = H_{2t}$$

$$B_{1n} = B_{2n} \tag{2.1.6}$$

where subscripts t and n denote the tangential and the normal components of field vectors, respectively.

In Gauss's law (equation (2.1.4)) for the electric field, σ_f is the density of free charge. For the problems in this book, we prefer to use Gauss's law in the form

$$\nabla \cdot \mathbf{E} = \sigma / \varepsilon_0 \tag{2.1.7}$$

where σ is the total charge density comprising both the free charges and the bound charges of dielectric polarization, and ε_0 [F m^{-1}] is free-space permittivity.

Next we shall show how, for a problem of determining the static electric field, Maxwell's equations reduce to a boundary-value problem of scalar potential. It is apparent from the vector identity $\nabla \times \nabla U = 0$ that if the curl of a vector vanishes, then the vector can be expressed as the gradient of a scalar potential function. It is apparent from Maxwell's equation (2.1.1) that for a static system, the curl of the electric field vanishes and the field is thus purely laminar, hence:

$$\mathbf{E} = -\nabla U \tag{2.1.8}$$

Consider the earth model depicted in Fig. 2.1. Surfaces S and A are physical boundaries which separate regions V_1, V_2 and V_3 from each other. The resistivities in V_1, V_2 and V_3 are ρ_1, ρ_2 and ρ_3 [Ω m], respectively. Surface B does not

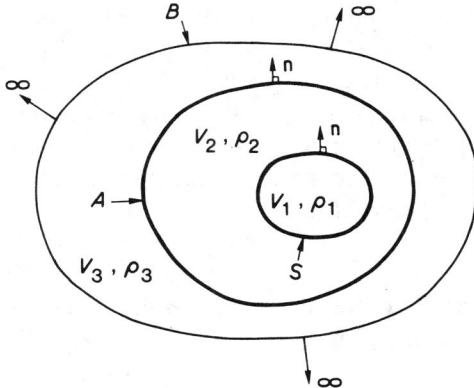

Figure 2.1 Earth model for electrical potential problem.

coincide with physical boundaries but is instead a fictitious surface used for illustrating the behaviour of potential at infinity. Resistivity ρ_1 in V_1 may be inhomogeneous, whereas ρ_2 and ρ_3 are assumed to be homogeneous.

In formulating the differential representation of U for this model, use is made of Ohm's law

$$\mathbf{E} = \rho \, \mathbf{j} \qquad (2.1.9)$$

as the relation between field strength $\mathbf{E}$ and current density $\mathbf{j}$, and the continuity relation for a static system, obtained from equation (2.1.5) as

$$\nabla \cdot \mathbf{j} = 0$$

Equations (2.1.8) and (2.1.9), together with equation (2.1.5), give for the potential:

$$\nabla \cdot \left(\frac{1}{\rho} \nabla U \right) = 0 \qquad (2.1.10)$$

which is the partial differential equation satisfied in the earth model under consideration. Resistivities ρ_2 and ρ_3 were assumed to be constant, and thus for regions V_2 and V_3, equation (2.1.10) simplifies to Laplace's equation

$$\nabla^2 U = 0 \qquad (2.1.11)$$

The general solutions of equations (2.1.10) and (2.1.11) must satisfy physically relevant homogeneous boundary conditions on surfaces S and A, and be regular on surface B as B recedes to infinity. It is easy to recognize that this problem can be solved analytically by the separation of variables only for geometrically very simple earth models. For a typical modelling problem in applied geophysics, the anomalous region is complicated and, consequently, more flexible methods must be used in the calculation of these models.

12 Electrostatics and potential theory

In accordance with the subject of this book, the boundary-value problem is converted into an integral equation by applying the Green's function technique. The sources of static electric field are the charges. Consequently, we reformulate the generalized Laplace equation (2.1.10) into Poisson's equation, with charges distributed in region V_1 appearing as the source term.

Taking the divergence on both sides of equation (2.1.8) and by substituting equation (2.1.7), we obtain for U:

$$\nabla^2 U = -\sigma/\varepsilon_0 \qquad (2.1.12)$$

Equation (2.1.12) is Poisson's equation for the electric potential. The unknown charge density σ represents both the volume charge in V_1 and the surface charge on S.

Since, according to Maxwell's equations, the tangential component of the electric field and the normal component of current density are continuous across physical boundaries, we require for potential U the following conditions on A:

$$U_3 = U_2$$
$$\frac{1}{\rho_3}\frac{\partial U_3}{\partial n} = \frac{1}{\rho_2}\frac{\partial U_2}{\partial n} \qquad (2.1.13)$$

In addition to equation (2.1.13), it is required as the regularity condition at infinity that RU is bounded as R approaches infinity.

In the following, the differential representation of potential U, defined by Poisson's equation (2.1.12) and boundary conditions (2.1.13), is converted into an integral representation by using the Green's function technique. The mathematical details are kept to a minimum here; for more rigorous treatment of the Green's function technique, the reader is referred to Roach [7].

2.2 INTEGRAL REPRESENTATION OF ELECTRICAL POTENTIAL

We shall now use the Green's function technique to convert the boundary-value problem for potential U, defined by equation (2.1.12) and boundary conditions (2.1.13), into integral representation of the form:

$$U(\mathbf{R}) = \frac{1}{\varepsilon_0} \int_V G(\mathbf{R}, \mathbf{R}_0)\, \sigma(\mathbf{R}_0)\, dV_0 \qquad (2.2.1)$$

where integration region V is supposed to consist of volume V_1 and boundary S of the model illustrated in Fig. 2.1. G is Green's function for equation (2.1.12) and the boundary conditions are as in equation (2.1.13). The superposition in equation (2.2.1) is possible because of the linearity of the system considered.

It is plausible to choose the same basic relations for Green's function as for the potential. Hence, G is to satisfy Poisson's equation

$$\nabla^2 G = -\delta(\mathbf{R} - \mathbf{R}_0) \tag{2.2.2}$$

where δ is Dirac's delta function and the boundary conditions are to be

$$G_3 = G_2 \tag{2.2.3}$$

$$\frac{1}{\rho_3} \frac{\partial G_3}{\partial n} = \frac{1}{\rho_2} \frac{\partial G_2}{\partial n}$$

on surface A. Further, G must be regular at infinity.

Green's function obtained from these conditions can be written as a sum of two terms:

$$G = G_0 + G_S \tag{2.2.4}$$

In equation (2.2.4) G_0 is the singular function:

$$G_0 = \frac{1}{4\pi |\mathbf{R} - \mathbf{R}_0|} \tag{2.2.5}$$

The function G_0 is a particular solution of equation (2.2.2). It is regular at infinity, and, together with its derivative, automatically continuous on A. G_0 is generally called the **whole-space Green's function**.

The function G_S is a non-singular function, which is the solution of Laplace's equation

$$\nabla^2 G_S = 0 \tag{2.2.6}$$

that satisfies boundary conditions (2.2.3) on surface A. *It represents the potential due to charges induced on boundaries of type A by the electrostatic interaction between source region V and boundaries A.*

The formal method of deducing an integral representation (2.2.1) from the differential equation and boundary conditions is based on the application of Green's second identity. Green's formulas are valid for two scalar functions which, together with their first and second derivatives, are continuous in a closed region and on its boundary. For a more complete presentation of Green's identities, see Kellogg [8]. Singularities at points and along lines can occur and the problem is then treated by suitably isolating the singularities.

Consider the region $V_2 + V_1$, enclosed by the closed surface A (Fig. 2.1). Assume that all anomalous charge is in region V_1, and that the physical boundaries of the surrounding space are represented by surface A. Let us substitute U and G into Green's second identity and apply it to the region $V_2 + V_1$ interior to boundary A:

$$\int_{V_1+V_2} (G_2 \nabla^2 U_2 - U_2 \nabla^2 G_2) \, dV = \int_A \left(G_2 \frac{\partial U_2}{\partial n} - U_2 \frac{\partial G_2}{\partial n} \right) dA \tag{2.2.7}$$

where $\mathbf{R}_0$ is in V_2 and $\mathbf{R}$ in $V_2 + V_1$, and the operations are performed with respect to variable $\mathbf{R}$. Substituting Poisson's equations (2.1.12) and (2.2.2) results in equation (2.2.7) acquiring the form:

$$U_2 = \frac{1}{\varepsilon_0} \int_{V_1} G_2 \sigma \, dV + \int_A \left(G_2 \frac{\partial U_2}{\partial n} - U_2 \frac{\partial G_2}{\partial n} \right) dA \qquad (2.2.8)$$

Applying Green's second identity to the region V_3 exterior to surface A and keeping point $\mathbf{R}_0$ in V_2, the volume integral on the left-hand side of Green's identity vanishes because U and G satisfy Laplace's equation in V_3. Further, the integral over the outermost surface B, which has been introduced in order to make the exterior region V_3 closed, approaches zero as B recedes to infinity. Green's second identity thus takes the form:

$$0 = -\int_A \left(G_3 \frac{\partial U_3}{\partial n} - U_3 \frac{\partial G_3}{\partial n} \right) dA \qquad (2.2.9)$$

Considering that U satisfies the boundary conditions of equation (2.1.13), it is now our aim, using the boundary conditions on G and U, to eliminate from equation (2.2.8) the unspecified values of U and its normal derivative on the boundary A. Multiplying equation (2.2.9) by ρ_2/ρ_3 and adding it to equation (2.2.8) we obtain, dropping the subscript on U,

$$U = \frac{1}{\varepsilon_0} \int_V G_2 \sigma \, dV$$

$$+ \int_A \left(G_2 \frac{\partial U_2}{\partial n} - U_2 \frac{\partial G_2}{\partial n} - \frac{\rho_2}{\rho_3} G_3 \frac{\partial U_3}{\partial n} + \frac{\rho_2}{\rho_3} U_3 \frac{\partial G_3}{\partial n} \right) dA \qquad (2.2.10)$$

Substitution of the boundary conditions (2.1.13) and (2.2.3) makes the surface integral in equation (2.2.10) vanish and, after interchanging variables $\mathbf{R}$ and $\mathbf{R}_0$ and using the reciprocity property of Green's function $(G(\mathbf{R}_0, \mathbf{R}) = G(\mathbf{R}, \mathbf{R}_0))$, the potential U is eventually represented by equation (2.2.1).

So far we have considered the potential problem in relation to the properties of the space exterior to the source region. The nature of the sources in static electrical and magnetic problems can often be specified directly on physical considerations. This is of considerable advantage because it allows deeper understanding of the physical properties of the model being considered. Models which cannot be formulated thus or which are not suitable for solution on the basis of physical formulation may, in some instances, be solved by the device of fictitious sources.

In the next four sections we shall consider in more detail the most common source types associated with electrical models. These are the primary current electrode, the volume and surface charge distributions, and the electrical double layer.

2.3 PRIMARY CURRENT ELECTRODE

In electrical surveying, the model is excited by primary current fed galvanically into the earth, although the self-potential method is an exception, being a self-exciting system.

The simplest means of current feeding is a point current electrode. Let a point electrode be earthed at point $\mathbf{R}_P$ in a medium of resistivity ρ. Charge Q [C] is created at the earthing point by a generator. The electric field due to charge Q drives a current I spreading from the electrode into the environment.

In searching for the relation between I and Q, use is made of Gauss's law (equation (2.1.7)) in integral form:

$$\oint_s \mathbf{E} \cdot \mathbf{n} \, dS = Q/\varepsilon_0 \qquad (2.3.1)$$

where S is a surface enclosing Q, and $\mathbf{n}$ is the outward normal unit vector on S.

To formulate a constitutive relation between field $\mathbf{E}$ and current density $\mathbf{j}$ use is made of Ohm's law:

$$\mathbf{E} = \rho \, \mathbf{j} \qquad (2.3.2)$$

For the problems considered in this book, the medium is assumed to be linear, which means that its resistivity ρ does not depend upon the current density. Also, unless indicated to the contrary, ρ is assumed to be isotropic, that is, independent of the direction of current. Substituting equation (2.3.2) into equation (2.3.1), the following relation between current strength I and charge Q at the earthing point is obtained:

$$Q = \varepsilon_0 \rho \oint_s \mathbf{j} \cdot \mathbf{n} \, dS = \varepsilon_0 \rho I \qquad (2.3.3)$$

The potential U at point $\mathbf{R}$ generated by charge Q at point $\mathbf{R}_p$ is:

$$U(\mathbf{R}) = \frac{1}{\varepsilon_0} G(\mathbf{R}, \mathbf{R}_P) \, Q(\mathbf{R}_P) \qquad (2.3.4)$$

where G is Green's function, as given in Section 2.2, for the space under consideration. Substituting equation (2.3.3) into equation (2.3.4), we obtain for the primary potential of a point current electrode:

$$U(\mathbf{R}) = \rho \, G(\mathbf{R}, \mathbf{R}_P) \, I(\mathbf{R}_P) \qquad (2.3.5)$$

The potential due to several point electrodes k is obtained from equation (2.3.5) by simple summation:

$$U(\mathbf{R}) = \sum_k \rho \, G(\mathbf{R}, \mathbf{R}_k) \, I(\mathbf{R}_k) \qquad (2.3.6)$$

Correspondingly, the potential generated by line current electrode L is calculated by integration:

16   *Electrostatics and potential theory*

$$U(\mathbf{R}) = \int_L \rho\, G(\mathbf{R}, \mathbf{R}_P)\, i(\mathbf{R}_P)\, dL_P \tag{2.3.7}$$

where i is the line density of current emitted by electrode L into the earth.

2.4 VOLUME CHARGE DISTRIBUTION

In regions of continuous change in resistivity, a volume charge density is created if the current has a component parallel to the resistivity gradient.

The relation between charge density σ and electric field $\mathbf{E}$ is represented by Gauss's law:

$$\nabla \cdot \mathbf{E} = \sigma/\varepsilon_0 \tag{2.4.1}$$

Substituting equation (2.3.2) into equation (2.4.1) and applying the conservation relation for electric current in a static system,

$$\nabla \cdot \mathbf{j} = 0 \tag{2.4.2}$$

we obtain the following relation between current density and the associated volume charge density:

$$\sigma = \varepsilon_0\, \mathbf{j} \cdot \nabla\rho \tag{2.4.3}$$

Consider a charge distributed in region V_1 with volume density $\sigma(\mathbf{R}_0)$ [C m^{-3}]. The potential at point $\mathbf{R}$ outside V_1 due to this charge is represented in accordance with Equation (2.2.1) by

$$U(\mathbf{R}) = \frac{1}{\varepsilon_0} \int_{V_1} G(\mathbf{R}, \mathbf{R}_0)\, \sigma(\mathbf{R}_0)\, dV_0 \tag{2.4.4}$$

When $\mathbf{R}$ is in V_1, U in equation (2.4.4) is defined as

$$U(\mathbf{R}) = \frac{1}{\varepsilon_0} \lim_{v \to 0} \int_{V_1 - v} G(\mathbf{R}, \mathbf{R}_0)\, \sigma(\mathbf{R}_0)\, dV_0 \tag{2.4.5}$$

where v is a small volume enclosing point $\mathbf{R}$. The integral is improper due to the singular part of Green's function (equation (2.2.5)), but the contribution of v vanishes as v approaches zero, and the integral is convergent and of the form of equation (2.4.4).

Similar considerations show that for the integral representation of field strength $\mathbf{E}$, the contribution of v also vanishes as v approaches zero, and the following integral is always convergent:

$$\mathbf{E}(\mathbf{R}) = -\frac{1}{\varepsilon_0} \int_{V_1} \sigma(\mathbf{R}_0)\, \nabla G(\mathbf{R}, \mathbf{R}_0)\, dV_0 \tag{2.4.6}$$

It is worth mentioning that for the Green's functions occurring in this book, the limit of the contribution from the small volume v in improper volume integrals is in general independent of the shape of v, or, alternatively, vanishes as v approaches zero in all its dimensions. The same applies to the small surface s in surface integrals, as will be seen in Sections 2.5 and 2.6. This is a useful property which facilitates the solution of integral equations.

2.5 SURFACE CHARGE DISTRIBUTION

We shall now see by taking an appropriate limit of Equation (2.4.3) that the source distribution associated with a discontinuity in resistivity is a layer of surface charge (Fig. 2.2). Consider a current passing through the boundary S between regions 1 and 2 having resistivity values ρ_1 and ρ_2. Let $\mathbf{n}$ be the unit normal vector of S defined as positive towards region 2.

Assume that a charge layer is distributed on S with surface density $\sigma_S(\mathbf{R}_0)$. This charge generates a potential $U(\mathbf{R})$ outside S:

$$U(\mathbf{R}) = \frac{1}{\varepsilon_0} \int_S G(\mathbf{R}, \mathbf{R}_0) \, \sigma_S(\mathbf{R}_0) \, dS_0 \tag{2.5.1}$$

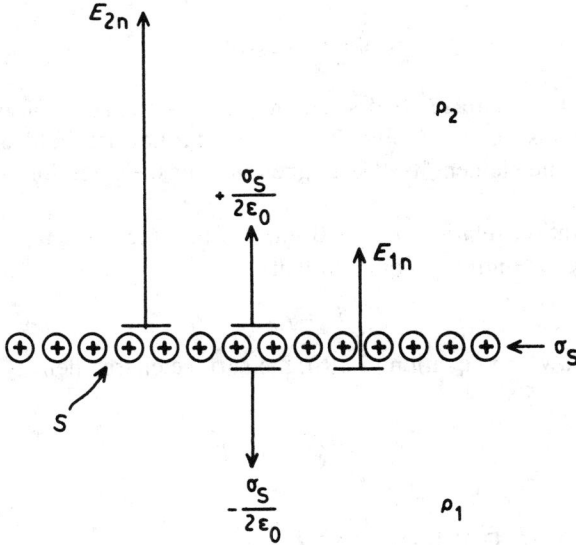

Figure 2.2 Behaviour of normal electrical field on surface charge distributed on resistivity discontinuity.

When $\mathbf{R}$ is on S, the potential U is obtained as the limit

$$U(\mathbf{R}) = \frac{1}{\varepsilon_0} \lim_{s \to 0} \int_{S-s} G(\mathbf{R}, \mathbf{R}_0) \, \sigma_S(\mathbf{R}_0) \, dS_0 \tag{2.5.2}$$

where s is a small element of S enclosing point $\mathbf{R}$. The contribution of s vanishes as s approaches zero. Thus the integral is convergent and U on S is expressed by equation (2.5.1).

The field strength at point $\mathbf{R}$ outside S is given by:

18   *Electrostatics and potential theory*

$$E(\mathbf{R}) = -\frac{1}{\varepsilon_0} \int_S \sigma_S(\mathbf{R}_0) \nabla G(\mathbf{R}, \mathbf{R}_0) \, dS_0 \qquad (2.5.3)$$

where the differentiation is performed with respect to $\mathbf{R}$, keeping $\mathbf{R}_0$ fixed.

Consider the limit of the normal component of field strength at points $\mathbf{R}_1$ and $\mathbf{R}_2$ of regions 1 and 2, as $\mathbf{R}_1$ and $\mathbf{R}_2$ approach point $\mathbf{R}$ on surface S along the common normal $\mathbf{n}$. Using the relation $E_n = -\partial U/\partial n$, we obtain

$$\lim_{\mathbf{R}_2 \to \mathbf{R}} E_n(\mathbf{R}_2) = \frac{\sigma_S(\mathbf{R})}{2\varepsilon_0} - \frac{1}{\varepsilon_0} \int_S \sigma_S(\mathbf{R}_0) \frac{\partial G(\mathbf{R}, \mathbf{R}_0)}{\partial n} \, dS_0$$

$$\lim_{\mathbf{R}_1 \to \mathbf{R}} E_n(\mathbf{R}_1) = -\frac{\sigma_S(\mathbf{R})}{2\varepsilon_0} - \frac{1}{\varepsilon_0} \int_S \sigma_S(\mathbf{R}_0) \frac{\partial G(\mathbf{R}, \mathbf{R}_0)}{\partial n} \, dS_0 \qquad (2.5.4)$$

where the surface integrals are convergent improper integrals.

It can be concluded from equation (2.5.4) that the electric field normal component has a discontinuity

$$E_{2n} - E_{1n} = \sigma_S/\varepsilon_0 \qquad (2.5.5)$$

on surface S. The jump in field strength across a surface charge is easy to understand by inspection of Fig. 2.2, where the normal field on a surface element due to the element itself is aligned in opposite directions on opposing sides of S.

The conservation relation div $\mathbf{j} = 0$ implies that the normal component of current density is continuous on S so that

$$j_{2n} = j_{1n} \qquad (2.5.6)$$

Using Ohm's law and equation (2.5.6), the surface charge density is obtained from equation (2.5.5) as:

$$\sigma_S = \varepsilon_0 (\rho_2 - \rho_1) j_n \qquad (2.5.7)$$

2.6 ELECTRICAL DOUBLE LAYER

In addition to the simple charge layer, the electrical double layer is also of great importance, consisting essentially of a layer of dipoles, distributed on surface S with dipole axes perpendicular to S (Fig. 2.3).

If regions 1 and 2 of space are not in direct galvanic contact with each other but are separated by surface impedance Z [Ω m^2] on S, the normal components of the field strength and the current density are continuous when passing through S, but the potential is discontinuous. In accordance with Ohm's law we obtain, for the difference in potential values of U_1 and U_2 on opposite sides of S:

$$U_2 - U_1 = -Zj_n \qquad (2.6.1)$$

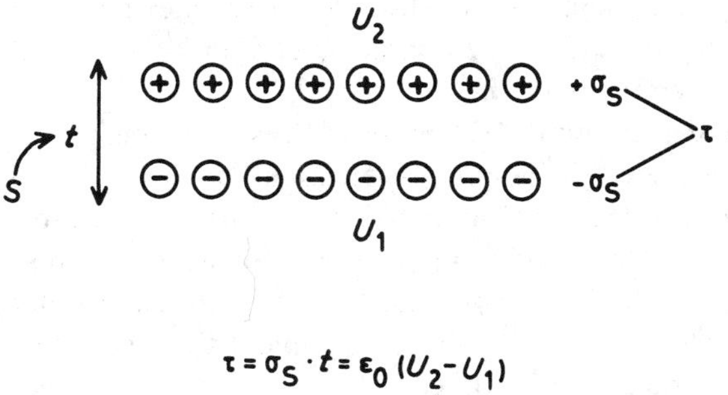

```
         U₂
    ⊕⊕⊕⊕⊕⊕⊕⊕  +σ_s
  ↗t                        ↘
 S  ↓                        τ
    ⊖⊖⊖⊖⊖⊖⊖⊖  -σ_s
         U₁
```

$$\tau = \sigma_s \cdot t = \varepsilon_0 (U_2 - U_1)$$

Figure 2.3 Behaviour of potential on an electrical double layer.

where **j** is the current density. The electrostatic source distribution associated with the discontinuity in potential is the electric double layer of moment density $\tau [\mathrm{C\,m^{-1}}]$.

Green's function for the potential generated by a double source is obtained from the Green's function G of the simple source by differentiating with respect to the source coordinate $\mathbf{R}_0$ in the direction of the dipole moment, that is, the normal $\mathbf{n}$ of surface S:

$$G'(\mathbf{R}, \mathbf{R}_0) = \frac{\partial G(\mathbf{R}, \mathbf{R}_0)}{\partial n_0} \tag{2.6.2}$$

The double layer generates potential U at point $\mathbf{R}$ outside S given by

$$U(\mathbf{R}) = \frac{1}{\varepsilon_0} \int_S G'(\mathbf{R}, \mathbf{R}_0)\, \tau(\mathbf{R}_0)\, \mathrm{d}S_0 \tag{2.6.3}$$

We shall now consider the limit of U in region 1 at point $\mathbf{R}_1$ and in region 2 at point $\mathbf{R}_2$ as $\mathbf{R}_1$ and $\mathbf{R}_2$ approach surface S at point $\mathbf{R}$ along the common normal $\mathbf{n}$:

$$\lim_{\mathbf{R}_2 \to \mathbf{R}} U(\mathbf{R}_2) = \frac{\tau(\mathbf{R})}{2\varepsilon_0} + \frac{1}{\varepsilon_0} \int_S G'(\mathbf{R}, \mathbf{R}_0)\, \tau(\mathbf{R}_0)\, \mathrm{d}S_0 \tag{2.6.4a}$$

$$\lim_{\mathbf{R}_1 \to \mathbf{R}} U(\mathbf{R}_1) = -\frac{\tau(\mathbf{R})}{2\varepsilon_0} + \frac{1}{\varepsilon_0} \int_S G'(\mathbf{R}, \mathbf{R}_0)\, \tau(\mathbf{R}_0)\, \mathrm{d}S_0 \tag{2.6.4b}$$

The surface integrals in equations (2.6.4) are improper but convergent.

It is seen from equations (2.6.4) that the potential generated by a double layer has a discontinuity

$$U_2 - U_1 = \tau/\varepsilon_0 \tag{2.6.5}$$

on surface S. Substituting equations (2.6.1) into equation (2.6.5) gives the relation between moment density τ and current density **j**:

$$\tau = \varepsilon_0 (U_2 - U_1) = -\varepsilon_0 Z j_n \tag{2.6.6}$$

The normal component of field strength approaches a common limit as the calculation point approaches the double layer from opposite sides of S, which means that E_n is continuous on S.

The relation in equation (2.6.1) applies to cases in which a current generates a physically real electric double layer. Double layers are also evoked as fictitious sources of electrical potential. Fictitious sources are useful in some instances in the formal solution of problems for which the physical formulation is not itself suitable for solution. However, in using fictitious sources a conceptual difficulty arises because the source distribution does not have any clear relation to the physical sources.

3

Electrical methods

3.1 INTRODUCTION

The aim of the preceding chapter was to summarize some aspects of electrostatics and potential theory which form an important basis for treating practical electrical modelling problems. In this chapter we shall apply the integral equation technique to solve some specific modelling problems that are encountered in the interpretation of electrical survey data.

Among electrical methods we include a variety of survey techniques which employ a direct current flowing in the earth. In most electrical methods, such as the resistivity method, the *mise-à-la-masse* method and the induced polarization (IP) method, the model is excited by a primary current supplied galvanically to the earth by an electrode system. An exception to this is the self-potential method in which the primary current is generated in a geological formation by spontaneous physical and chemical processes.

A common feature of electrical methods is their theoretical framework, which is based on the static electric potential theory. The IP method is distinguished from other electrical methods by its exploitation of the frequency dependence of resistivity. It employs a time-dependent source signal, but the space dependence of the response signal can be described with a fair degree of accuracy by means of potential theory, which involves the exclusion of electromagnetic induction from the theory. Such methods are generally called **relaxation** methods. Potential-theory treatment implies that the frequency dependence of the IP signal is due exclusively to the frequency dependence of the resistivity. This is to be compared with the complete formulation of time-varying electromagnetic fields, in which the frequency dependence is included in Green's function as a fundamental property of Maxwell's equations.

The integral equation technique is particularly suitable for solving static problems, because the physical characteristics of static problems are often sufficiently simple to be directly and illustratively represented by integral equation formulation. Furthermore, static models are generally well behaved in the numerical solution of integral equations.

3.2 RESISTIVITY OF ROCKS

The resistivity of rocks is a widely varying property, ranging from $10^{-6}\,\Omega$ m for graphite and pyrrhotite to $10^{12}\,\Omega$ m for quartzitic rocks [9].

The principal minerals present in rocks are generally insulators. In a rock composed of insulating minerals, current is carried by aqueous ions through water occupying the pores, fissures and fractures. As the degree of pore permeability and the content and concentration of electrolytes increase, the resistivity of the rock decreases.

Important exploration targets include such semi-conducting minerals as graphite, pyrrhotite, chalcopyrite, galena and magnetite, all of which are electronic conductors commonly having a resistivity two orders of magnitude lower than that of the electrolyte in the pore spaces. The contribution of these minerals to the rock conductivity (the reciprocal of resistivity) depends significantly on their texture and the mode in which they are disseminated through the rock. For example, pyrite, chalcopyrite and pyrrhotite can form an interconnected network which makes the rock highly conductive, whereas galena and magnetite usually occur as isolated individual grains. Galena and magnetite ores are thus usually less conductive than, for example, pyrite ores.

The bulk resistivity of rocks containing disseminated ore minerals is complex and depends upon the frequency of the current. This property, called **dispersion**, is utilized in the induced polarization method. If the ionic paths in rock pores are impeded by intervening semi-conducting mineral grains, a two-phase system is formed where current is carried alternately by electrons and ions. At the phase boundaries between the mineral grains and the electrolyte, the current carriers are interchanged by electrochemical oxidation–reduction reactions, ion diffusion and capacitive effects. These processes are referred to as **electrode polarization**, the macroscopic outcome of which is the complex surface impedance on phase boundaries. This impedance makes the resistivity of rock complex, with the result that the current and the electric field oscillate in different phases and the magnitude of the resistivity decreases with increasing frequency.

Dispersion of resistivity is also observed in the absence of electronically conducting minerals. Clay particles cause so-called **membrane polarization**, where unbalanced charges in clays attract a diffuse cloud of positive ions. These ions block pore passages and the blocking effect grows stronger with decreasing frequency of current. Again, the resistivity of the rock is complex and its magnitude decreases with increasing frequency.

An additional property associated with polarizable rocks is the nonlinearity of resistivity. At low current densitites – that is, $10^{-3}\,\mathrm{Am^{-2}}$ or less – which are typical of resistivity and induced polarization surveys, the resistivity is independent of current density, and the relation between electric field and current density is linear. If considerably higher current density is applied, say, 10–100 A m^{-2}, new electrochemical reactions start to decrease the impedance at phase boundaries. These reactions can be used to characterize the mineralization and the electrolyte, or the ion-exchanging mineral system of membrane polarization.

The nonlinearity of resistivity is used in the identification of mineralization types and in interpretation of the geometries of mineralizations. This application is known as the **contact polarization** method [10].

In this book only methods operating in the linearity range are considered.

3.3 RESISTIVITY METHOD

3.3.1 Introduction

In applying resistivity methods, the geological formations under investigation are energized by current introduced into the ground through current electrodes and the potential generated by the current is measured by potential electrodes. The apparent resistivity, which is a quantity commonly used as a basis for interpretation, is then calculated by dividing the measured potential values by the geometrical factor characteristic of the electrode system and by the current. The geometrical factor can be regarded as Green's function for the electrode array used in the measurement. The apparent resistivity is defined in such a way that it equals the true resistivity if the measurement were performed on homogeneous ground.

As the resistivity system is linear, the solution for a model excited by a particular current electrode array can be obtained from the solution for a single point current electrode by superposition. Hence, in the following theoretical considerations, the primary potential is taken to have an unspecific form, U_P. For each particular application, the specific form of primary potential is then susbtituted for U_P.

Regarding electrical modelling, the solution for the resistivity problem also contains in principle the solutions for the *mise-à-la-masse* and induced polarization problems. In fact, the *mise-à-la-masse* method can be considered to be one of the resistivity methods having a special electrode array. On the other hand, the induced polarization method is based on the behaviour of earth resistivity as a function of the frequency of primary current. Following common practice, however, we shall consider these two methods in separate sections.

In this section we shall consider the 3-dimensional integral equations for charge density and for potential distributed on discontinuity surfaces of resistivity. The special cases of perfectly conductive bodies and of very long or very thin bodies are also considered.

3.3.2 Integral equations for charge density

General charge distribution

We shall now consider the resistivity problem for an earth model illustrated in Fig. 2.1. Region V_1 represents the anomalous body and regions V_2 and V_3 the surrounding space. The anomalous resistivity ρ_1 may be varying from point to point, but ρ_2 and ρ_3 are homogeneous. The homogeneity of ρ_2 should be understood symbolically; it represents in a general sense the resistivity distribution in the space surrounding the anomalous region, including such structures as the layered earth. The problem is treated on a physical basis by considering the electric potential as being due to surface charge of density σ_S distributed on the resistivity discontinuity S, and volume charge of density σ distributed in a manner corresponding to the continuous variations of resistivity ρ_1 in region V_1. Let us simplify the notation by substitution $q_S = \sigma_S/\varepsilon_0$ and $q = \sigma/\varepsilon_0$. The potential U at point $\mathbf{R}$ of region V_2 can now be represented in accordance with equations (2.5.1) and (2.4.4) as

$$U(\mathbf{R}) = U_P(\mathbf{R}) + \int_S G(\mathbf{R}, \mathbf{R}_0)\, q_S(\mathbf{R}_0)\, dS_0 + \int_{V_1} G(\mathbf{R}, \mathbf{R}_0)\, q(\mathbf{R}_0)\, dV_0 \qquad (3.3.1)$$

where $\mathbf{R}_0$ is the source point located in region V_1 or on surface S, and G is Green's function for the space surrounding anomalous region $S + V_1$, as defined in Section 2.2. U_P is the primary potential.

Generally in geophysical modelling, the charge densities q_S and q are unknown. The next step in solving the problem is to develop a form for equation (3.3.1) such that an integral equation can be formed in order to solve the unknown charge densities q_S and q. Accordingly, the calculation point $\mathbf{R}$ is moved into the source region $S + V_1$, and charge densities q_S and q are substituted for potential U on the left-hand side of equation (3.3.1).

Consider first the case when the calculation point $\mathbf{R}$ is on surface S. In substituting q_S for U, use is made of equation (2.5.7):

$$q_S = (\rho_2 - \rho_1) j_n$$

which, applied together with Ohm's law, for example on the negative side V_1 of S,

$$j_n = \frac{E_n}{\rho_1} = -\frac{1}{\rho_1} \frac{\partial U}{\partial n}$$

gives for the normal derivative of U:

$$\frac{\partial U}{\partial n} = -\frac{\rho_1 q_S}{(\rho_2 - \rho_1)} \qquad (3.3.2)$$

Taking into consideration the singularity of Green's function in the source region in accordance with equation (2.5.4), we get from equation (3.3.1), after differentiation with respect to coordinate **R** and substitution from equation (3.3.2):

$$\frac{\rho_1 q_s}{(\rho_2-\rho_1)} = -\frac{\partial U_P}{\partial n} - \frac{q_s}{2} - \int_S \frac{\partial G}{\partial n} q_s \, dS_0 - \int_{V_1} \frac{\partial G}{\partial n} q \, dV_0 \qquad (3.3.3)$$

on surface S. The second term on the right-hand side of equation (3.3.3) is due to the singularity which has been removed from the surface integral by a limit process.

When the calculation point **R** is in region V_1, charge density q can be substituted for U by using equation (2.4.3) in V_1:

$$q = \mathbf{j} \cdot \nabla \rho_1$$

Denoting the unit vector in the direction of $\nabla \rho_1$ by **a** and applying Ohm's law, we obtain from the above formula:

$$\frac{\partial U}{\partial a} = -\frac{\rho_1 q}{|\nabla \rho_1|} \qquad (3.3.4)$$

As stated in Section 2.4, the singularity of Green's function in a volume charge region does not introduce any additional term to the field strength. After differentiation at point **R** in direction a, and substitution from equation (3.3.4) we get from equation (3.3.1):

$$\frac{\rho_1 q}{|\nabla \rho_1|} = -\frac{\partial U_P}{\partial a} - \int_S \frac{\partial G}{\partial a} q_s \, dS_0 - \int_{V_1} \frac{\partial G}{\partial a} q \, dV_0 \qquad (3.3.5)$$

in region V_1. The final Fredholm integral equation of the second kind can now be written for simultaneously solving the surface charge density q_s on S and volume charge density q in V_1:

$$\frac{\rho_2+\rho_1}{2(\rho_2-\rho_1)} q_s(\mathbf{R}) = -\frac{\partial U_P(\mathbf{R})}{\partial n} - \int_S \frac{\partial G(\mathbf{R}, \mathbf{R}_0)}{\partial n} q_s(\mathbf{R}_0) \, dS_0$$

$$- \int_{V_1} \frac{\partial G(\mathbf{R}, \mathbf{R}_0)}{\partial n} q(\mathbf{R}_0) \, dV_0 \qquad \mathbf{R} \text{ is on } S$$

$$\frac{\rho_1}{|\nabla \rho_1|} q(\mathbf{R}) = -\frac{\partial U_P(\mathbf{R})}{\partial a} - \int_S \frac{\partial G(\mathbf{R}, \mathbf{R}_0)}{\partial a} q_s(\mathbf{R}_0) \, dS_0$$

$$- \int_{V_1} \frac{\partial G(\mathbf{R}, \mathbf{R}_0)}{\partial a} q(\mathbf{R}_0) \, dV_0 \qquad \mathbf{R} \text{ is in } V_1$$

(3.3.6)

Integral equation (3.3.6), which is divided into two parts – one required to be satisfied on surface S and the other in region V_1 – can be discretized and solved numerically for q_s and q, as described in Section 1.3. As soon as q_s and q have been solved, the potential U can be calculated from equation (3.3.1) for arbitrary points of regions V_1 and V_2.

26 Electrical methods

The field strength can be calculated from the relation $\mathbf{E} = -\nabla U$ by the formula:

$$\mathbf{E}(\mathbf{R}) = -\nabla U_P(\mathbf{R}) - \int_S \nabla G(\mathbf{R}, \mathbf{R}_0) \, q_s(\mathbf{R}_0) \, dS_0 - \int_{V_1} \nabla G(\mathbf{R}, \mathbf{R}_0) \, q(\mathbf{R}_0) \, dV_0 \quad (3.3.7)$$

The solution domain of integral equation (3.3.6) consists of both surface S and volume V_1, which requires a considerable effort in obtaining the numerical solution. If the shape and inner structure of region V_1 are very complicated, it might be more practical to solve the problem by the finite element or by finite difference techniques rather than by the integral equation technique.

Surface charge distribution

The application of the integral equation technique is most advantageous in the solution of relatively simple earth models. In particular, if the resistivity ρ_1 of an anomalous body V_1 can be considered to be homogeneous, the volume integral term in equation (3.3.1) vanishes and the secondary potential is represented by a surface integral. Correspondingly, integral equation (3.3.6) reduces to a surface integral equation. In the following we shall consider this simple model, for which resistivities ρ_1 and ρ_2 differ from each other but are both homogeneous. The formulation is performed for one anomalous body but it can be immediately generalized so as to apply to models containing several anomalous bodies.

For homogeneous ρ_1 and ρ_2, potential U can be represented in the following form:

$$U(\mathbf{R}) = U_P(\mathbf{R}) + \int_S G(\mathbf{R}, \mathbf{R}_0) \, q_s(\mathbf{R}_0) \, dS_0 \quad (3.3.8)$$

(see Fig. 2.1) where q_s is the unknown surface charge density on S. The surface charge is negative on parts of S where current passes through S from a poorer conductive medium to a better one, and positive in the opposite case. For a typical situation in a resistivity survey where the primary current sources are located in the outer region V_2 of S, the total surface charge Q_S on S is zero:

$$Q_S = \int_S q_s \, dS = 0$$

In the so-called **weak scattering problems**, for which the contrast between resistivities ρ_1 and ρ_2 is weak, various simple approximation methods can be used for determining q_s. Methods analogous to the application of demagnetization factor techniques in magnetic modelling may be mentioned as an example. In contrast, at an interface S with a strong resistivity contrast, the secondary field on S due to its own charge distribution is significant in comparison with the primary field, making the internal interaction of the surface charge distribution strong. Consequently, more sophisticated methods must be used to obtain q_s when the contrast between ρ_1 and ρ_2 is high.

To form an integral equation for q_s, we differentiate both sides of equation (3.3.8) with respect to coordinate **R** and move the calculation point **R** to surface S. Taking into consideration the singularity of Green's function on S similarly as in equation (3.3.3), and substituting q_s for the normal derivative of U according to equation (3.3.2), we obtain:

$$\frac{\rho_2+\rho_1}{2(\rho_2-\rho_1)} q_s(\mathbf{R}) = -\frac{\partial U_P(\mathbf{R})}{\partial n} - \int_S \frac{\partial G(\mathbf{R}, \mathbf{R}_0)}{\partial n} q_s(\mathbf{R}_0) \, dS_0 \qquad (3.3.9)$$

After a numerical solution for q_s has been obtained, the potential U can be calculated at arbitrary points of regions V_1 and V_2 by equation (3.3.8).

Equation (3.3.8) and the associated integral equation (3.3.9) are widely applied in electrical modelling. The computer programs for their numerical solution mostly utilize the point-matching method with pulse functions as the subsectional basis, as described in Section 1.3.3. Let us now demonstrate this common case.

Surface S is divided into N areal subsections s_j, $j = 1, \ldots, N$, and q_s is assumed to be constant within each s_j. This discretization transforms equation (3.3.9) into the following form:

$$\frac{\rho_2+\rho_1}{2(\rho_2-\rho_1)} q_s(\mathbf{R}_i) = -\frac{\partial U_P(\mathbf{R}_i)}{\partial n_i} - \sum_{i=1}^{N} \frac{\partial G_{ij}(\mathbf{R}_i, \mathbf{R}_j)}{\partial n_i} q_s(\mathbf{R}_j) \qquad (3.3.10)$$

where i and j take on all values $1, \ldots, N$. G_{ij} is Green's function which represents the potential at point $\mathbf{R}_i$ due to surface charge distributed on subsection s_j:

$$G_{ij}(\mathbf{R}_i, \mathbf{R}_j) = \int_{S_j} G(\mathbf{R}_i, \mathbf{R}_0) \, dS_0 \qquad (3.3.11)$$

equivalently to equation (1.3.7). Note that the singular part of integral (3.3.11), when $i = j$, must be considered as a principal value, that is, the singular point $\mathbf{R}_0 = \mathbf{R}_i$ is excluded from the integral. The contribution of this point is included in the left-hand side of equation (3.3.9) corresponding to the second term on the right-hand side of equation (3.3.3).

An important question associated with the numerical solution is how fast it converges towards the correct solution of the problem as the dimensions of the subsections decrease. Schulz [11] has shown that if surface S is divided into boundary elements, each of which is a square of length h, the difference between the calculated potential U_N and the exact potential U is

$$|U_N - U| \leq ch$$

where c is a constant which depends upon the continuity of U and cannot generally be determined more exactly. Since the number of boundary elements increases as the square of the element density along the linear dimensions, the convergence is seen to be rather slow.

28 *Electrical methods*

As an application of equations (3.3.8) and (3.3.9), consider the apparent resistivity of an inclined three-dimensional prism embedded in a two-layered half-space. The resistivity of the prism is 25 Ω m and that of the environment 5000 Ω m. The resistivity of the surface layer, when the layer is present, is 500 Ω m. Green's function for the two-layered earth is given in Appendix A. Figure 3.1(a) illustrates the effect of dip on apparent resistivity obtained by a dipole–dipole array, from which it is apparent that the dipole–dipole array is only weakly sensitive to dip. From Fig. 3.1(b) we notice that an overburden layer ten times as conductive as the underlying half-space considerably reduces the intensity of the apparent resistivity anomaly.

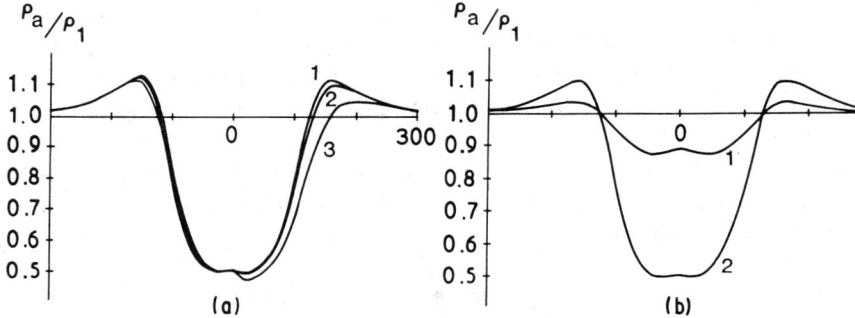

Figure 3.1 Apparent resistivity of three-dimensional prism in a two-layered half-space. Dipole-dipole array, $a = 50$ m, $N = 4$. Depth and thickness of prism 50 m, depth extent 260 m and strike length 150 m. (a) Dip 90° (1), 75° (2) and 45° (3), surface layer not present. (b) Dip 90°, surface layer present (1) and not present (2). After [12].

As another application of equation (3.3.9) consider the effect of electrostatic interaction between two anomalous bodies. Figure 3.2 shows apparent resistivity anomalies for a model consisting of two prisms. The solid line depicts the correct situation where the interaction between the prisms has been taken into account, and the hatched line represents the numerically easier but physically incorrect situation where the interaction has been disregarded. In the former case, the integration domain in equation (3.3.9) is the total anomalous surface composed of two prism surfaces. In the latter case, equation (3.3.9) is solved separately for both prisms and the total potential is obtained as the sum of the two potentials. We see that the interaction between the prisms is moderate when the resistivity contrast is 5. When the contrast increases to 50, the interaction is so strong that quite incorrect modelling results will be obtained if the interaction is not incorporated into the calculations. This effect is physically a result of the chage distributions on adjacent faces of different prisms strengthening each other due to the coupling effect. The positive maximum of the anomaly is due to the strong positive electric field between the prisms, which is caused by these charges.

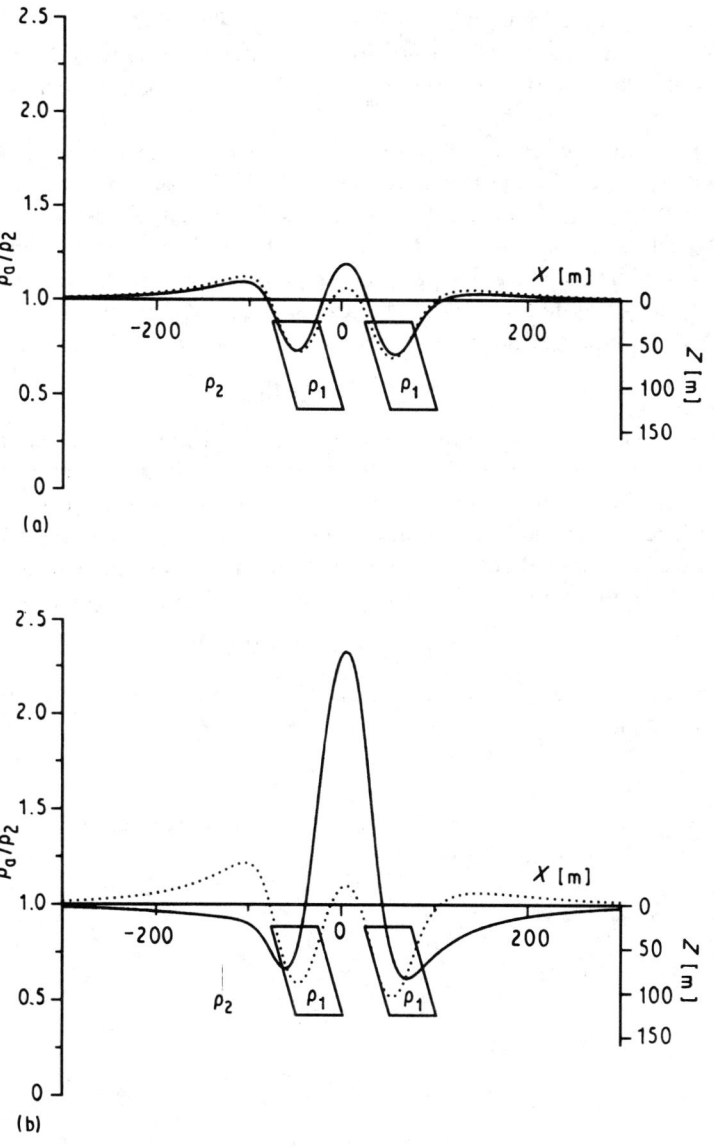

Figure 3.2 Apparent resistivity of a system of two prisms in homogeneous half-space, with electrostatic interaction between prisms taken into account (solid curve) and disregarded (broken curve). Strike length 100 m, dip 75°. Mid-gradient configuration with current electrodes at $x = \pm 1000$ m (a) $\rho_2/\rho_1 = 5$; (b) $\rho_2/\rho_1 = 50$.

3.3.3 Integral equations for potential

In Section 3.3.2 an integral equation solution was given for the resistivity problem, which was based on true physical source distributions. Next we shall introduce an alternative formulation in which the secondary potential is generated by a fictitious source distribution [13].

To formulate the resistivity problem in terms of a boundary-value problem, consider the earth model illustrated in Fig. 2.1. Surfaces S and A divide the space into regions V_1, V_2 and V_3 which have homogeneous resistivities ρ_1, ρ_2 and ρ_3. The fictitious surface B represents the limit of the space at infinity. Surface S is the anomalous source region and A represents the physical boundaries of the surrounding space. The primary current electrodes are located in region V_2. The partial differential equation satisfied by current density $\mathbf{j}$ is

$$\nabla \cdot \mathbf{j} = I\delta(\mathbf{R} - \mathbf{R}_P) \tag{3.3.12}$$

where δ is Dirac's delta function and $\mathbf{R}_P$ is the location of a point current electrode emitting current I into the ground. Solutions for multi-electrode arrays can be obtained by simple superposition. Applying Ohm's law, $\mathbf{E} = \rho \mathbf{j}$, and taking into consideration the fact that the electric field $\mathbf{E}$ can be represented as the gradient of a scalar potential, $\mathbf{E} = -\nabla U$, the partial differential equation governing the resistivity problem is Poisson's equation:

$$\nabla^2 U = -\rho_2 I\delta(\mathbf{R} - \mathbf{R}_P) \tag{3.3.13}$$

which is valid in region V_2. In V_1 and V_3, equation (3.3.13) simplifies to Laplace's equation:

$$\nabla^2 U = 0 \tag{3.3.14}$$

The solutions of equations (3.3.13) and (3.3.14) must satisfy the following conditions on interfaces S and A:

$$U_1 = U_2$$
$$\frac{1}{\rho_1}\frac{\partial U_1}{\partial n} = \frac{1}{\rho_2}\frac{\partial U_2}{\partial n}, \quad \text{on} \quad S \tag{3.3.15}$$

and

$$U_2 = U_3$$
$$\frac{1}{\rho_2}\frac{\partial U_2}{\partial n} = \frac{1}{\rho_3}\frac{\partial U_3}{\partial n}, \quad \text{on} \quad A \tag{3.3.16}$$

In addition to conditions (3.3.15) and (3.3.16), RU must be bounded, as R approaches infinity.

We shall now convert the boundary-value problem defined by equations (3.3.13)–(3.3.16) into an integral equation. Green's function G for the model considered satisfies Poisson's equation

Resistivity method 31

$$\nabla^2 G(\mathbf{R}, \mathbf{R}_0) = -\delta(\mathbf{R} - \mathbf{R}_0) \quad (3.3.17)$$

the solution to which must satisfy the boundary conditions

$$G_2 = G_3 \quad (3.3.18)$$

$$\frac{1}{\rho_2}\frac{\partial G_2}{\partial n} = \frac{1}{\rho_3}\frac{\partial G_3}{\partial n}, \quad \text{on} \quad A$$

as well as the regularity condition, which is that RG is bounded as R approaches infinity.

Let us substitute functions U and G into Green's second identity and apply the identity first to region V_2 of Fig. 2.1 assuming that both $\mathbf{R}$ and $\mathbf{R}_0$ are in V_2:

$$\int_{V_2}(G_2\nabla^2 U_2 - U_2\nabla^2 G_2)\,dV = \int_S\left(-G_2\frac{\partial U_2}{\partial n} + U_2\frac{\partial G_2}{\partial n}\right)dS$$

$$+ \int_A\left(G_2\frac{\partial U_2}{\partial n} - U_2\frac{\partial G_2}{\partial n}\right)dA \quad (3.3.19)$$

The differentiations and integrations in equation (3.3.19) are performed with respect to variable $\mathbf{R}$.

Applying Green's second identity to region V_3 and keeping point $\mathbf{R}_0$ in V_2, it can be demonstrated using arguments similar to those in Section 2.2 (formulas (2.2.9) and (2.2.10)), that the surface integral in equation (3.3.19) over A vanishes, provided that boundary conditions (3.3.16) and (3.3.18) are satisfied on A. Substituting equations (3.3.13) and (3.3.17) and dropping the subscript from G, identity (3.3.19) takes the form:

$$-U_P(\mathbf{R}_0) + U_2(\mathbf{R}_0) = \int_S\left[-G(\mathbf{R}, \mathbf{R}_0)\frac{\partial U_2(\mathbf{R})}{\partial n} + U_2(\mathbf{R})\frac{\partial G(\mathbf{R}, \mathbf{R}_0)}{\partial n}\right]dS \quad (3.3.20)$$

where the first term stands for the primary potential,

$$U_P(\mathbf{R}_0) = \rho_2 G(\mathbf{R}_P, \mathbf{R}_0)\,I\,(\mathbf{R}_P) \quad (3.3.21)$$

We next apply Green's second identity to region V_1 keeping point $\mathbf{R}_0$ in region V_2. As the primary sources and the singular points of Green's function are outside V_1, the volume integral on the left-hand side of the identity vanishes and thus:

$$0 = \int_S\left[G(\mathbf{R}, \mathbf{R}_0)\frac{\partial U_1(\mathbf{R})}{\partial n} - U_1(\mathbf{R})\frac{\partial G(\mathbf{R}, \mathbf{R}_0)}{\partial n}\right]dS \quad (3.3.22)$$

In equation (3.3.20), the secondary potential in V_2 is represented by a surface integral, where the first term represents the potential of simple source and the second term the potential of a double source. Developement of the problem on the basis of the simple source term would lead to the same physical solution as was obtained in Section 3.3.2. Now we shall adopt the alternative view and form the integral representation of the potential as being due to a double source. To eliminate the simple source term in equation (3.3.20), we multiply equation (3.3.22) by ρ_2/ρ_1 and add the equation thus obtained to equation (3.3.20). Recalling that $U_1 = U_2 = U$ on surface S (boundary condition (3.3.15)), replacing $\mathbf{R}_0$ with $\mathbf{R}$, and using the reciprocity relation $G(\mathbf{R}_0, \mathbf{R}) = G(\mathbf{R}, \mathbf{R}_0)$, we obtain for potential U in region V_2:

$$U(\mathbf{R}) = U_p(\mathbf{R}) + \frac{\rho_1 - \rho_2}{\rho_1} \int_S G'(\mathbf{R}, \mathbf{R}_0) U(\mathbf{R}_0) \, dS_0 \qquad \mathbf{R} \text{ in } V_2 \qquad (3.3.23)$$

where $G'(\mathbf{R}, \mathbf{R}_0)$ is given by equation (2.6.2). The primary potential U_p is obtained from equation (3.3.21):

$$U_P(\mathbf{R}) = \rho_2 G(\mathbf{R}, \mathbf{R}_P) I(\mathbf{R}_P) \qquad (3.3.24)$$

To derive the representation of U in region V_1, point $\mathbf{R}_0$ is located in V_1 when applying Green's second identity to regions V_1 and V_2. A consideration similar to that above yields the equation

$$U(\mathbf{R}) = \frac{\rho_1}{\rho_2} U_P(\mathbf{R}) + \frac{\rho_1 - \rho_2}{\rho_2} \int_S G'(\mathbf{R}, \mathbf{R}_0) U(\mathbf{R}_0) \, dS_0 \qquad \mathbf{R} \text{ in } V_1 \qquad (3.3.25)$$

where U_P is given by equation (3.3.24).

In the physically based integral equation solution considered in Section 3.3.2, the potential was represented by one and the same equation in both region V_1 and V_2. In the formulation based on fictitious double sources, potential U has dissimilar representations on different sides of S. In addition, the resistivity value appearing in the primary potential in region V_1 is ρ_1, instead of the true resistivity value ρ_2. This difference in primary potentials for regions V_1 and V_2 compensates for the discontinuity in Green's function $G'(\mathbf{R}, \mathbf{R}_0)$ on surface S, thus making the total potential continuous on S, as required by boundary condition (3.3.15).

To find the potential U on S, we develop equation (3.3.23) into an integral equation by moving the calculation point $\mathbf{R}$ in region V_2 to surface S. Taking into consideration the singularity of Green's function in accordance with equation (2.6.4a) and by substituting the term caused by the singularity $(\rho_1 - \rho_2)U/\rho_1$ for τ/ε_0, we obtain the intergral equation for U:

$$\frac{\rho_1 + \rho_2}{2\rho_1} U(\mathbf{R}) = U_P(\mathbf{R}) + \frac{\rho_1 - \rho_2}{\rho_1} \int_S G'(\mathbf{R}, \mathbf{R}_0) U(\mathbf{R}_0) \, dS_0 \qquad (3.3.26)$$

Equation (3.3.26) is solved for U using numerical methods. The known potential values on surface S are then substituted into equations (3.3.23) and (3.3.25) for calculating U at arbitrary points of regions V_2 and V_1.

Note that exactly the same integral equation (3.3.26) would have been obtained if the derivation had started from equation (3.3.25).

Every constant potential is a solution to the homogeneous counterpart of equation (3.3.26) (that is, $U_P = 0$), when ρ_1 tends to zero [11]. Then, according to the Fredholm theorems, the nonhomogeneous equation (3.3.26) has no unique solutions. Thus, the formulation given above cannot be used in solving potential problems for models containing perfect conductors. On the other hand, Eloranta [13] has demonstrated that equation (3.3.26) works well up to resistivity contrasts as high as $\rho_2/\rho_1 = 10^4$, beyond which the model has probably reached electrostatic saturation. This means that the formulation presented is valid for a fairly wide range of resistivity contrasts.

The integral formulation given above can be characterized as the representation of the potential by analytical continuation of the potential of fictitious sources, to be distinguished from the physically based formulation given in the preceding section. The relative merits of these two formulations depend upon the aims of the modelling. If, for example, one is examining the physical properties of the model, then the physical representation is more appropriate. Otherwise, it is more a matter of how the formulations behave when solved numerically. In conventional resistivity modelling the two formulations nevertheless are practically equivalent. The applicability of the formulations to *mise-à-la-masse* modelling is considered in Section 3.5.

3.3.4 Integral equation for perfect conductor model

When the relative resistivity contrast between the anomalous body and the environment increases, the electrostatic state of the charge distributed on the surface of the body becomes saturated. In the saturated state, further change of the resistivity contrast does not change the charge distribution. If the anomalous body is more conductive than the envirnment, the potential of the body becomes constant once the system becomes saturated. The resistivity contrast at which the potential can be considered to be constant depends upon the shape of the body. Bodies for which all the three dimensions are approximately equal reach the equipotential state at a lower resistivity contrast than more elongated bodies. When all the dimensions of the body are approximately of the same length, the saturation state is reached at a restivity contrast of $\rho_2/\rho_1 = 100$. In the following we shall consider the application of the equipotential condition to the integral equation solution of resistivity problems.

The earth model considered is shown in Fig. 2.1, where S represents the surface of anomalous body V_1 and A the boundaries of the space surrounding the body. Surface B represents the limit of the space at infinity. The current electrodes are located in region V_2. Resistivity ρ_1 is assumed to be so small in relation to ρ_2 that the potential U in body V_1 and on its boundary S can be considered constant. The potential in V_2 can be written in the form

$$U(\mathbf{R}) = U_P(\mathbf{R}) + \int_S G(\mathbf{R}, \mathbf{R}_0) q_S(\mathbf{R}_0) \, dS_0 \tag{3.3.27}$$

where $q_s = \sigma_s/\varepsilon_0$, and σ_s is the charge density on S. Primary potential U_P is given by equation (3.3.24), and Green's function satisfies the usual conditions given by equations (3.3.17) and (3.3.18).

To solve the unknown charge density q_s we develop equation (3.3.27) into an integral equation by moving the calculation point $\mathbf{R}$ on to surface S. Considering that the singularity of Green's function on S does not introduce any additional term, as explained in Section 2.5, and that the potential is constant on S, $U = U_C$, we obtain

$$U_C = U_P(\mathbf{R}) + \int_S G(\mathbf{R}, \mathbf{R}_0) q_s(\mathbf{R}_0) dS_0 \qquad \mathbf{R} \text{ and } \mathbf{R}_0 \text{ on } S \qquad (3.3.28)$$

Since in addition to q_s, U_C is also unknown, equation (3.3.28) is underdetermined and one supplementary equation is needed for solving q_s, U_C. We get this equation from the conservation relation of charge, by writing equation (2.3.1) in the form

$$\frac{Q}{\varepsilon_0} = \int_{S(I)} q_s \, dS = \rho I \qquad (3.3.29)$$

where I is the current emitted inside $S(I)$. For surfaces $S(\text{no } I)$ which do not enclose current electrodes, we obtain:

$$\int_{S(\text{no } I)} q_s \, dS = 0 \qquad (3.3.30)$$

The unknown surface charge density q_s, and also the potential U_C, can now be solved from equation (3.3.28) in combination with equation (3.3.30). If the model is composed of several anomalous bodies V_{1i} with resistivity values ρ_i, surface S is also composed of several separated partial surfaces S_i. For this case, equation (3.3.28) must be satisfied simultaneously on the entire surface S, whereas each subsurface S_i achieves its own constant potential U_{Ci}. Correspondingly, equation (3.3.30) is satisfied separately on each partial surface S_i. The surface charge density q_s obtained by the solution is substituted into equation (3.3.27) to calculate the potential U at any point of region V_2.

The numerical solution of equations (3.3.28) and (3.3.30) gives a fair degree of accuracy with sparse discretization. The application of this method is advantageous whenever the resistivity contrast on surface S is sufficiently high to make the assumption of constant potential valid.

Consider as an application the apparent resistivity anomalies of a cube located in a homogeneous half-space, obtained by the three different integral equation formulations given above (see Fig. 3.3). The conductivity contrast between the cube and the environment is 500, which is sufficiently high to make the potential in the cube constant, and thus the anomalies calculated from integral equations (3.3.9) and (3.3.26) are comparable with the anomaly obtained from equations (3.3.28) and (3.3.30).

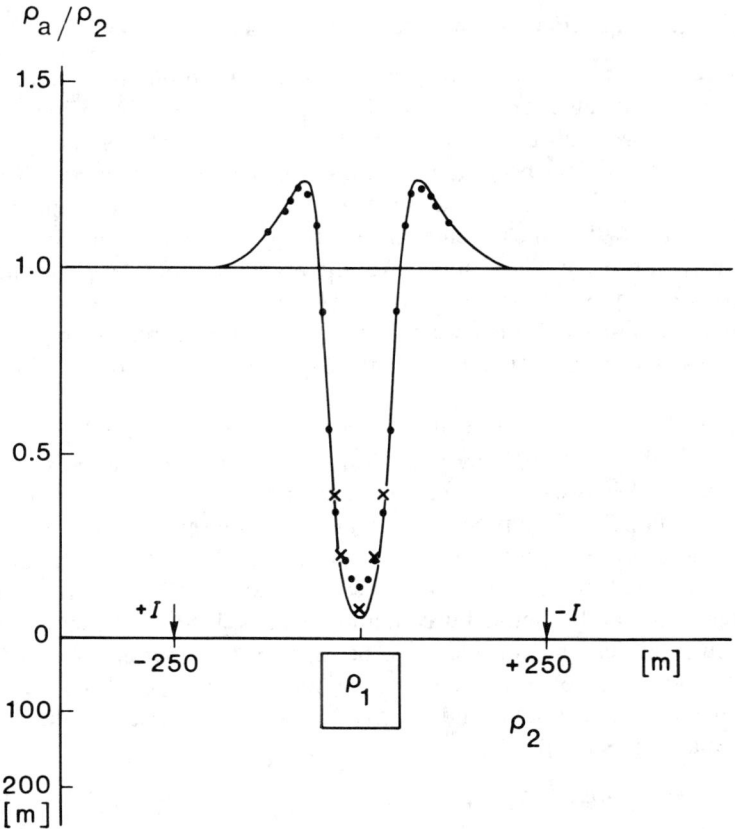

Figure 3.3 Apparent resistivity of a cube located in homogeneous half-space calculated from integral equations of surface charge (·), of potential (×), and of surface charge using equipotential condition (solid curve). Gradient configuration, $\rho_1 = 1\ \Omega\,\text{m}$, $\rho_2 = 500\ \Omega\,\text{m}$. After [13].

The numerical solution was obtained by dividing the surface of the cube into uniform square-shaped elements and by assuming that the source density is constant within each element. In the solution of the integral equation (3.3.9) for surface charge, each face was divided into 144 elements, and in the solution of the two other equations, into 49 elements. Considering that the anomaly amplitude obtained by the discretization used increases with the number of elements, it may be noted that the solution to integral equation (3.3.9) (marked by dots) needs the highest number of elements. An apparent explanation for this is that integral equation (3.3.9) has been deduced from the field representation, whereas the two others have been deduced from the potential one. Therefore, the product of Green's function and the unknown source density varies more rapidly in equation (3.3.9) than in the two comparable equations. Nevertheless, all three curves are fairly compatible with each other.

3.3.5 Integral equations for two-dimensional earth models

Geological structures are often elongated, persisting in one horizontal direction to a considerable extent in comparison with other dimensions of the model. Direct application of three-dimensional modelling technique for these structures is inefficient because the number of boundary elements increases with increasing strike length, and thus makes the set of algebraic equations to be solved large. When measurements have been made using long line current electrodes aligned parallel to the strike, the modelling may be performed by applying a purely two-dimensional model. However, point current electrodes are commonly used in field surveys, which makes the problems three-dimensional, in which case purely two-dimensional model calculations are inappropriate.

In the following, we shall consider the integral equation modelling of the resistivity problem by applying the $2\frac{1}{2}$-dimensional model, that is, where the anomalous body is two-dimensional but the current electrode system is three-dimensional [14, 15]. The problem is treated by representing the model functions by their Fourier transforms in the direction of the strike. The original three-dimensional problem is thus converted into a set of two-dimensional problems for properly sampled wavenumber values. If the strike length of the anomalous body can be considered to be infinite, the Fourier integral transform is applied. If the strike length of the body is finite but still sufficiently long to make the effect of the end faces on the potential negligible, the Fourier series expansion is applied.

Infinite strike length

Consider an earth model in which ρ_1 is the resistivity of a two-dimensional anomalous body with the strike parallel to the y-axis, and ρ_2 is the resistivity of the environment of the body. In this case a single anomalous body is being considered but models containing several bodies can also be directly generalized.

Let us start with the three-dimensional representation of potential given by equation (3.3.8):

$$U(\mathbf{R}) = U_P(\mathbf{R}) + \int_S G(\mathbf{R}, \mathbf{R}_0)\, q_S(\mathbf{R}_0)\, dS_0 \qquad (3.3.31)$$

where q_S is the charge density distributed on surface S of the body. The three-dimensional integral equation deduced from equation (3.3.31) for solving q_S is written here for convenience from equation (3.3.9):

$$\frac{\rho_2+\rho_1}{2(\rho_2-\rho_1)} q_S(\mathbf{R}) = -\frac{\partial U_P(\mathbf{R})}{\partial n} - \int_S \frac{\partial G(\mathbf{R}, \mathbf{R}_0)}{\partial n} q_S(\mathbf{R}_0)\, dS_0 \qquad (3.3.32)$$

We denote the difference between the three-dimensional position vectors $\mathbf{R}$ and $\mathbf{R}_0$ as follows:

$$|\mathbf{R} - \mathbf{R}_0| = [|\mathbf{r} - \mathbf{r}_0|^2 + (y - y_0)^2]^{1/2}$$

$$|\mathbf{r} - \mathbf{r}_0| = [(x - x_0)^2 + (z - z_0)^2]^{1/2} \tag{3.3.33}$$

The three-dimensional Green's function can now be written in the form

$$G(\mathbf{R}, \mathbf{R}_0) = G(\mathbf{r}, \mathbf{r}_0; y, y_0),$$

where $\mathbf{r}$ and $\mathbf{r}_0$ are the two-dimensional components of vectors $\mathbf{R}$ and $\mathbf{R}_0$ in the plane $y = 0$. Equation (3.3.31) can now be written in the form

$$U(\mathbf{r}; y) = U_P(\mathbf{r}; y) + \oint_C \int_{-\infty}^{\infty} G(\mathbf{r}, \mathbf{r}_0; y, y_0)\, q_S(\mathbf{r}_0; y_0)\, dy_0\, dC_0 \tag{3.3.34}$$

where C is the contour of the cross-section of surface S in the plane $y = 0$. The double integral consists of the line integral along contour C of the convolution of functions G and q_S along the y-axis.

Define the Fourier transform pair as

$$F\{f(y)\} = \overline{f}(k_y) = \int_{-\infty}^{\infty} f(y)\, e^{-ik_y y}\, dy \tag{3.3.35a}$$

$$F^{-1}\{\overline{f}(k_y)\} = f(y) = \frac{1}{2\pi} \int_{-\infty}^{\infty} \overline{f}(k_y)\, e^{ik_y y}\, dk_y \tag{3.3.35b}$$

where k_y is the wavenumber and $i^2 = -1$. Now we apply equation (3.3.35a) to obtain the Fourier transform of both sides of equation (3.3.34). Taking into consideration that the Fourier transform of a convolution of two functions equals the product of the Fourier transforms of these functions, we obtain:

$$\overline{U}(\mathbf{r}; k_y) = \overline{U}_P(\mathbf{r}; k_y) + \oint_C \overline{G}(\mathbf{r}, \mathbf{r}_0; k_y)\, \overline{q}_S(\mathbf{r}_0; k_y)\, dC_0 \tag{3.3.36}$$

where $\overline{q}_S(k_y)$ is the unknown spectral value of charge density q_S at wavenumber k_y.

To solve $\overline{q}_S$, the Fourier transform is taken on both sides of equation (3.3.32), resulting in

$$\frac{\rho_2 + \rho_1}{2(\rho_2 - \rho_1)} \overline{q}_S(\mathbf{r}; k_y) = -\frac{\partial \overline{U}_p(\mathbf{r}; k_y)}{\partial n}$$

$$-\oint_C \frac{\partial \overline{G}(\mathbf{r}, \mathbf{r}_0; k_y)}{\partial n} \overline{q}_S(\mathbf{r}_0; k_y)\, dC_0 \tag{3.3.37}$$

where both $\mathbf{r}$ and $\mathbf{r}_0$ are on contour C. The integral in equation (3.3.37) must be understood as a principal value, that is, the singular point $\mathbf{r} = \mathbf{r}_0$ is excluded from the integral and its contribution is taken into consideration on the left-hand side of the equation.

Equation (3.3.37) is solved for a set of wavenumbers k_y which adequately sample the spectrum $\bar{q}_S$ so that the numerical inverse Fourier transform of the spectrum can be made. The $\bar{q}_S$ values thus solved are substituted into equation (3.3.36) which can then be used to calculate the spectral values of the potential $\bar{U}(\mathbf{r}; k_y)$ for each sampled k_y. The potential $U(\mathbf{R})$ at an arbitrary point $\mathbf{R}$ of the space is finally obtained as the inverse Fourier transform equation (3.3.35b) of $\bar{U}$:

$$U(\mathbf{R}) = F^{-1}\{\bar{U}(\mathbf{r}; k_y)\} \tag{3.3.38}$$

An important matter is the choice of the wavenumbers in sampling the spectra. We shall describe after Snyder [14] the main factors related to the sampling of wavenumbers when the space surrounding the two-dimensional anomalous body is a homogeneous half-space. Since the potential spectrum due to a line element of C is the superposition of the spectrum of Green's function, sampling relations are thus strongly dependent upon the behaviour of Green's function. Consider the Fourier transform of the three-dimensional half-space Green's function as given in equation (1.1.5):

$$G(\mathbf{R}, \mathbf{R}_0) = \frac{1}{4\pi}\left(\frac{1}{|\mathbf{R} - \mathbf{R}_0|} + \frac{1}{|\mathbf{R} - \mathbf{R}_0'|}\right) \tag{3.3.39}$$

where $\mathbf{R}_0'$ is the mirror image of the source point $\mathbf{R}_0$ in relation to the ground surface $z = 0$.

The Fourier transforms of the Green's function and its normal derivative are as follows:

$$\bar{G}(\mathbf{r}, \mathbf{r}_0; k_y) = F\{G(\mathbf{R}, \mathbf{R}_0)\} = K_0(|\mathbf{r} - \mathbf{r}_0|k_y) + K_0(|\mathbf{r} - \mathbf{r}_0'|k_y) \tag{3.3.40}$$

$$\frac{\partial \bar{G}(\mathbf{r}, \mathbf{r}_0; k_y)}{\partial n} = \frac{\partial}{\partial n} F\{G(\mathbf{R}, \mathbf{R}_0)\} = \frac{(\mathbf{r} - \mathbf{r}_0) \cdot \mathbf{n}}{|\mathbf{r} - \mathbf{r}_0|^2} |\mathbf{r} - \mathbf{r}_0|k_y K_1(|\mathbf{r} - \mathbf{r}_0|k_y)$$

$$+ \frac{(\mathbf{r} - \mathbf{r}_0') \cdot \mathbf{n}}{|\mathbf{r} - \mathbf{r}_0'|^2} |\mathbf{r} - \mathbf{r}_0'|k_y K_1(|\mathbf{r} - \mathbf{r}_0'|k_y) \tag{3.3.41}$$

where K_0 and K_1 are modified Bessel functions of orders 0 and 1, respectively.

Equation (3.3.36) shows that the secondary potential is a superposition of Green's function (3.3.40). Taking into consideration the behaviour of function K_0 we can recognize that the high wavenumber contributions come from the Green's function at small distances between the source point and the calculation point. Therefore, the wavenumber at which the sampling of the spectrum can be truncated is determined by the calculation point and the source point for which the distance $|\mathbf{r} - \mathbf{r}_0|$ is a minimum, $r_{\min}$. The sampling of the spectra should thus be designed to adequately sample the spectrum of $K_0(r_{\min})$. This guarantees that all other contributions to the spectrum, which have less abundant high-frequency components, are also adequately sampled.

To determine the truncation point for k_y, it is first necessary to determine the value of argument rk_y above which function K_0 is negligible. We consider that spectrum $\overline{G}$ has been sufficiently attenuated at $r_{min} k_y = 4$, which corresponds to the value $K_0(4) = 0.01$. Hence, using the criterion $k_{y\,max} = r_{min}/4$, the appropriate wavenumber interval can then be obtained for each particular problem. Note that the case $k_y = 0$ represents the pure two-dimensional model.

If the potential has to be calculated for $y \neq 0$, wavelengths that are equal to or greater than $2y_{max}$ must be sampled. This requires that the first k_y to be sampled must be $k_{y\,min} \leq \pi / y_{max}$.

If the method of subsections is used in the numerical treatement of equations (3.3.36) and (3.3.37), the subsections are simply short line segments. Instead of using the conventional point-matching in solving integral equation (3.3.37), Snyder [14] uses the Galerkin method where Green's function is integrated along both the source and the calculation subsections. By this procedure, the quantity obtained by solving the discretized integral equation is the spectrum of the total charge on each subsection.

Finite strike length

Consider now that the strike length, L, of the model considered in the previous section is finite, but sufficiently long to make the contribution of the end faces on the potential negligible. The basic representation of potential is given by equation (3.3.31), and the integral equation for solving the charge density is given by equation (3.3.32), where S denotes the surface of the truncated cylindrical anomalous bodies. Let the plane $y = 0$ be the central normal plane of the bodies and consider only the symmetrical case, where the current electrodes are located in this plane. By substituting equation (3.3.33), equation (3.3.31) can be written in the form

$$U(\mathbf{r}; y) = U_P(\mathbf{r}; y) + \oint_C \int_{-L/2}^{L/2} G(\mathbf{r}, \mathbf{r}_0; y, y_0)\, q_S(\mathbf{r}_0; y_0)\, dy_0\, dC_0 \quad (3.3.42)$$

where functions U, G and q_S are defined only for $-L/2 \leq y \leq L/2$. The Fourier transform pair for the finite interval of y is defined as follows:

$$F\{f(y)\} = \overline{f}(k_y = 0) = \frac{1}{L} \int_{-L/2}^{L/2} f(y)\, dy \qquad k_y = 0$$

$$(3.3.43a)$$

$$F\{f(y)\} = \overline{f}(k_y) = \frac{2}{L} \int_{-L/2}^{L/2} f(y) \cos \frac{2\pi k_y y}{L}\, dy, \qquad k_y \neq 0$$

$$F^{-1}\{\overline{f}(k_y)\} = f(y) = \sum_{k_y = 0}^{N} \overline{f}(k_y) \cos \frac{2\pi k_y y}{L} \quad (3.3.43b)$$

The definition of transforms (3.3.43) as pure cosine transforms is possible due to the symmetry of the model about the plane $y = 0$. As can be seen from equation (3.3.43b), the spectra are now discrete, that is, the wavenumber k_y is an integer. The case $k_y = 0$ represents the truncated two-dimensional model where the current electrodes are line electrodes of length L emitting current of constant density I/L.

The truncation point N of the Fourier series must be chosen so that the spectra of functions G and q_S are adequately sampled. Earlier in this section we represented the sampling rules for the Fourier spectra of infinitely long bodies in terms of the argument $k_y r$. The corresponding quantity for the discrete spectrum of finite bodies is $2\pi k_y r/L$. Numerical calculations have shown that r should be considered here as the distance of the nearest current electrode from the anomalous body. If then ratio r/L is, for example, $\frac{1}{4}$, the Fourier series equation (3.3.43b) can be truncated at $N = 5$.

Representing G and q_S in equation (3.3.42) by their Fourier series expansions according to equation (3.3.43b) and also taking into consideration the orthogonality properties of sine and cosine functions, we obtain by direct calculation from equation (3.3.42):

$$\overline{U}(\mathbf{r}; k_y) = \overline{U}_P(\mathbf{r}; k_y) + \oint_C \overline{G}(\mathbf{r}, \mathbf{r}_0; k_y) \, \overline{q}_S(\mathbf{r}_0; k_y) \, dC_0 \qquad (3.3.44)$$

which represents the spectral value of potential at integer wavenumber value k_y. Functions $\overline{U}$, $\overline{U}_P$ and $\overline{q}_S$ are obtained from equation (3.3.43b). Green's function $\overline{G}$ is obtained from the Fourier transform of G as follows:

$$\overline{G} = L \cdot F\{G\} \qquad k_y = 0 \qquad (3.3.45)$$

$$\overline{G} = \frac{L}{2} \cdot F\{G\} \qquad k_y \neq 0$$

The integral equation for solving the spectral value of charge density $\overline{q}_S$ at wavenumber k_y takes the form of equation (3.3.37), which is solved for a sufficient number of wavenumber values $k_y = 0, 1, \ldots, N$. The solved values $\overline{q}_S$ are substituted into equation (3.3.44) which enables the calculation of the spectral values of potential at an arbitrary point $\mathbf{r}$ on the plane $y = 0$. Finally, the potential $U(\mathbf{R})$ at point $\mathbf{R}$ in the region $-L/2 \leq y \leq L/2$ can be calculated by summing the spectral values according to equation (3.3.43b):

$$U(\mathbf{r}; y) = \sum_{k_y=0}^{N} \overline{U}(\mathbf{r}; k_y) \cos \frac{2\pi k_y y}{L} \qquad (3.3.46)$$

As an application, let us consider the convergence of numerical solution towards the correct solution as the number of line subsections on the $2\frac{1}{2}$-dimensional anomalous body increases. We shall use the point-matching method for two different sets of basis functions for the charge density, namely pulse functions, and first-degree functions which cause the charge density to vary along the subsections in a linear fashion. Since the domain of the integral equation is a line, the information concerning the linear variation of charge density can be easily inserted into the system through Green's function [15, 16]. These two different subsectional bases give different kinds of approximation for the unknown charge density. For example, the shape of the charge density is often concave in the vicinity of corner points, in which case the pulse function approximation gives an underestimate of the charge density, whereas the linear approximation gives an overestimate. This makes it possible to find certain lower and upper limits for the correct solution, when the size of subsections is decreased in a series of numerical solutions using both of these bases.

The $2\frac{1}{2}$-dimensional model consists of a body of $100\,\text{m} \times 100\,\text{m}$ cross-section, located in a homogeneous half-space. Depth to the upper surface is 20 m and the strike length 1000 m. The conductivity of the body is ten times that of the environment and the electrode array is the mid-gradient configuration with a current electrode separation of 2000 m.

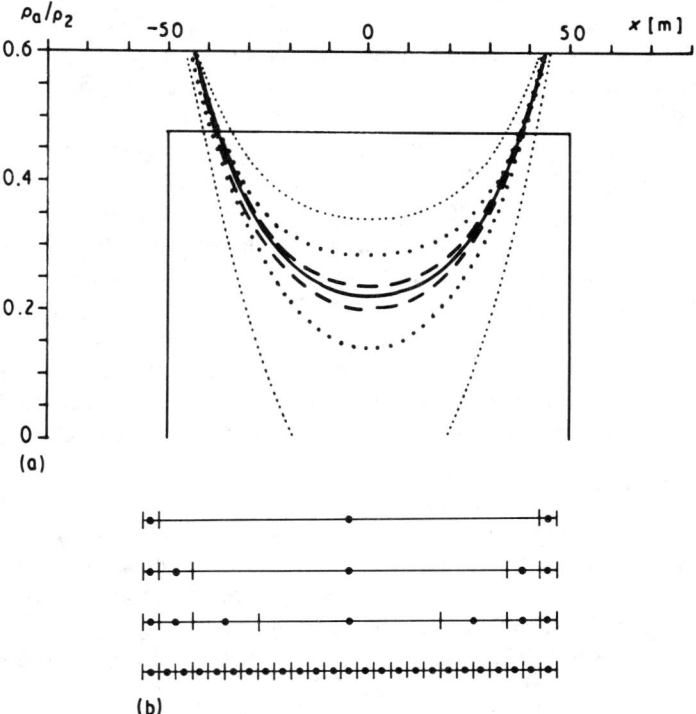

Figure 3.4 $2\frac{1}{2}$-dimensional model for an anomalous body with square cross-section located in a half-space. (a) Apparent resistivity calculated by using four different divisions into subsections. (b) Subsection divisions used in (a). After [17].

42 *Electrical methods*

Figure 3.4(a) illustrates the apparent resistivity of the model calculated for four different subsection densities. Each side of the square cross-section has been divided into subsections in the way depicted in Fig. 3.4(b). The curves shown are obtained by using both the pulse functions and the first-degree functions as basis functions. We notice that the anomalies obtained by using the first-degree basis functions are stronger than those obtained by using the pulse functions, but also that the curves associated with different basis functions approach each other as the density of subsections increases. At the highest density used in these calculations, that is, 25 subsections for each side, the anomalies are practically coincident, which suggests that we have obtained a solution which is very close to the correct anomaly curve.

3.3.6 Integrodifferential equations for thin conductor models

We shall now consider the integral equation formulation for a thin conductor model. In certain geological formations, such as mineralized veins and fracture zones, one dimension is generally much shorter than the others.

It is known that during electrical surveys, the contributions due to conductivity and thickness of a thin conductor cannot be determined separately, but rather that the response is controlled by a combined quantity, known as **longitudinal conductance**, which is equal to the product of conductivity and thickness.

The direct application of three-dimensional integral equations for solving potential problems of thin conductors is not practical because the equation must be solved on each boundary face of the conductor, and consequently requires a large number of subsections for the numerical solution. Moreover, the physical character of current channeling by longitudinal flow within the conductor is not adequately represented by three-dimensional formulation. Instead, an easier and more illustrative solution is obtained by considering the conductor to be so thin that its adjacent faces are joined, forming a simple conducting surface. In the following we shall derive this reduced model from the general three-dimensional integral representation for surface charge.

Solution for the potential

Let a thin conductor be embedded in a space whose Green's function is known. The resistivity of the conductor is ρ_1 and that of the medium surrounding it ρ_2 (Fig. 3.5(a)). When electric current is introduced into the system, current passing through the conductor generates on its boundaries surface charges of density σ_S, as given in equation (2.5.7). Our aim is now to make the conductor so thin that its adjacent boundary faces can be considered to be joined. As a result the conductor is represented by a simple two-dimensional surface A, bearing a surface charge of density σ_A. Let σ denote the total surface charge density consisting of σ_A and of charge σ_C distributed on the thin edges C of the conductor. In the following formulation we use charge density in the form $q = \sigma/\varepsilon_0$.

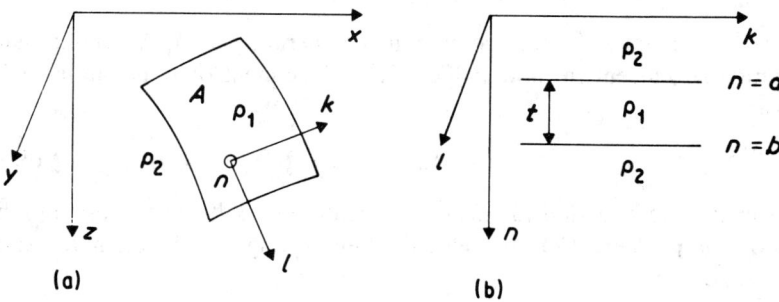

Figure 3.5 Thin conductor model. (a) General view. Resistivity of conductor ρ_1 and of bounding medium ρ_2. (b) Enlargement of part of the section perpendicular to conductor A in (a). Coordinate system (k, l, n) fixed to the conductor.

The potential $U(\mathbf{R})$ generated by the system can be written in the following form [18]:

$$U(\mathbf{R}) = U_P(\mathbf{R}) + \int_A G(\mathbf{R}, \mathbf{R}_0)\, q(\mathbf{R}_0)\, dA_0 \qquad (3.3.47)$$

where U_P is the primary potential and G is Green's function for the space surrounding conductor A. The core of the problem is now to determine the unknown charge density q in equation (3.3.47).

The charge density q is composed of q_A formed by joining the charges distributed on adjacent boundary faces of the conductor, and of q_C distributed on the thin edges. Consider first charge q_A.

The current flowing through boundary surfaces $n = a$ and $n = b$ (Fig. 3.5(b)) generates surface charges in accordance with equation (2.5.7):

$$q_a = (\rho_1 - \rho_2)\, j_{na} \qquad (3.3.48)$$

$$q_b = (\rho_2 - \rho_1)\, j_{nb}$$

where j_{na} and j_{nb} are the n-components of current density on surfaces $n = a$ and $n = b$

Let ρ_1 and the thickness of the conductor, $t = b - a$, be vanishingly small in such a way that the ratio t/ρ_1 is finite. In the limit, charges q_a and q_b form a simple charge layer distributed on surface A with density q_A. Taking into consideration that $\rho_1 \ll \rho_2$ we obtain:

$$q_A = q_a + q_b = \rho_2 (j_{nb} - j_{na}) \qquad (3.3.49)$$

In order to represent a_A in terms of the longitudinal component of current density in place of its normal component, we apply the continuity equation of current, $\nabla \cdot \mathbf{j} = 0$, within conductor A, thus

44 *Electrical methods*

$$\int_a^b \left(\frac{\partial j_n}{\partial n}\right) dn = -\int_a^b \left(\frac{\partial j_k}{\partial k} + \frac{\partial j_l}{\partial l}\right) dn = -\int_a^b \nabla_A \cdot \mathbf{j}\, dn \tag{3.3.50}$$

where $\nabla_A \cdot$ denotes the surface divergence operating on A. Assuming that the longitudinal current in the conductor is independent of n, equation (3.3.50) becomes

$$j_{nb} - j_{na} = -t\, \nabla_A \cdot \mathbf{j} \tag{3.3.51}$$

By substituting equation (3.3.51) into equation (3.3.49) and applying Ohm's law, $\mathbf{E} = \rho_1 \mathbf{j}$, where $\mathbf{E}$ is the electric field strength, we obtain for surface charge q_A:

$$q_A = -\rho_2 \nabla_A \cdot g\mathbf{E} \tag{3.3.52}$$

where g is the longitudinal conductance of the concuctor, defined by

$$g = t/\rho_1 \tag{3.3.53}$$

By analogy with equation (3.3.48), the charge density q_C on the thin edges C of the conductor is

$$q_C = (\rho_2 - \rho_1) j_C \tag{3.3.54}$$

where j_C is the component of the longitudinal current density vector (j_k, j_l) perpendicular to edge C (positive outwards from A). Since the thickness of the conductor is small in relation to the other dimensions of the model, the charge distributed on the edges can be considered as a line charge with line density Q_C:

$$Q_C = q_C\, t \tag{3.3.55}$$

By applying Ohm's law, $E_C = \rho_1 j_C$, equations (3.3.54) and (3.3.55) yield

$$Q_C = \rho_2 g E_C \tag{3.3.56}$$

where E_C is the component of the longitudinal electric field (E_k, E_l) perpendicular to edge C.

Considering that $q = q_A + q_C$ and using equation (3.3.55), equation (3.3.47) can be written in the form

$$U = U_P + \int_A G q_A \, dA_0 + \oint_C G Q_C \, dc_0 \tag{3.3.57}$$

where c denotes the line reduced from edges C. By substituting equations (3.3.52) and (3.3.56), we obtain from equation (3.3.57):

$$U = U_P + \rho_2 \left(\int_A -G\nabla_A \cdot g\, \mathbf{E}\, dA_0 + \oint_C G g E_C\, dc_0 \right) \tag{3.3.58}$$

By applying the partial differentiation rule to the first integrand in equation (3.3.58):

$$-G\nabla_A \cdot g\,\mathbf{E} = -\nabla_A \cdot Gg\,\mathbf{E} + g\,\mathbf{E}\cdot\nabla_A G$$

and Gauss's theorem to the first term on the right-hand side of the above formula:

$$\int_A -\nabla_A \cdot G\,g\mathbf{E}\,dA_0 = -\oint_c Gg\,\mathbf{E}_C\,dc_0$$

and substituting

$$(E_k, E_l) = -\nabla_A U \qquad (3.3.59)$$

equation (3.3.58) is transformed into the integrodifferential representation

$$U(\mathbf{R}) = U_P(\mathbf{R}) + \rho_2 \int_A -g(\mathbf{R}_0)\,\nabla_{A0}U(\mathbf{R}_0)\cdot\nabla_{A0}G(\mathbf{R},\mathbf{R}_0)\,dA_0 \quad (3.3.60)$$

where ∇_{A0} refers to the differentiation on surface A with respect to the source coordinates (k_0, l_0).

To find the potential U on surface A, equation (3.3.60) is developed into an integral equation by moving the calculation point $\mathbf{R}$ to A. The contribution of the small area containing the singular point $\mathbf{R} = \mathbf{R}_0$ vanishes as the area approaches zero and thus the integral equation is of the form of equation (3.3.60). After U has been solved on A, the potential can in principle be calculated at any point in the surrounding space by again applying equation (3.3.60). As shall be explained later, an algorithm developed at the Geological Survey of Finland, which uses equation (3.3.60) to calculate U outside conductor A, is only sufficiently accurate for moderate values of longitudinal conductance g; for high values of g, the results are inaccurate. A generally applicable algorithm for calculating potential outside A can nevertheless be obtained by a method in which the potential obtained on A from equation (3.3.60) is substituted into the left-hand side of equation (3.3.47), enabling the charge density q to be solved from the equation thus obtained. When q is known, the potential U can be calculated outside A by again applying equation (3.3.47).

As the product $\rho_2 g$ increases, the charge distribution on A approaches electrostatic saturation, after which the potential on A remains constant. This limiting case of perfect conductance can be treated by the procedure given above. However, it can also be solved simply by analogy with Section 3.3.4 by making the left-hand side of equation (3.3.47) constant, U_C in A. Since the numerical value of U_C is unknown, we need one supplementary equation for solving q, and for this we can use the conservation relation of charge:

$$\int_A q\,dA = 0 \qquad (3.3.61a)$$

when the current electrode is outside the surface A, and

46   *Electrical methods*

$$\int_A q \, dA = p_2 I \tag{3.3.61b}$$

when an electrode emitting current I is in A.

If surface A is composed of several separate subsurfaces A_i, the integration area in equation (3.3.47) is the entire surface A, whereas each subsurface A_i is at its own constant potential U_{Ci}. Correspondingly, equation (3.3.61a) or (3.3.61b) must be satisfied separately for each subsurface A_i.

Numerical solution

Equation (3.3.60) can be solved numerically, for example by using the point-matching method with pulse functions as a subsectional basis. Hence, the potential is assumed to be constant within each subsection and equation (3.3.60) is assumed to hold at the centre of each subsection. Partial derivatives in the integral are replaced by finite-difference relations in which central differences are used at interior points and forward or backward differences at edge points. The resulting set of algebraic equations is then solved by Gaussian elimination. The potential U is obtained as the solution at the centres of the subsections. The potential values solved for the subsections are substituted into the left-hand side of the discretized form of equation (3.3.47), from which the charge density q is then solved. The potential U is then calculated outside A by again applying equation (3.3.47).

As an application of the above formulation, let us consider the response of a thin square conductor to the resistivity measurement made by a gradient array. The conductor is situated in a homogeneous half-space and its longitudinal conductance g is varied. For the numerical solution, the conductor is

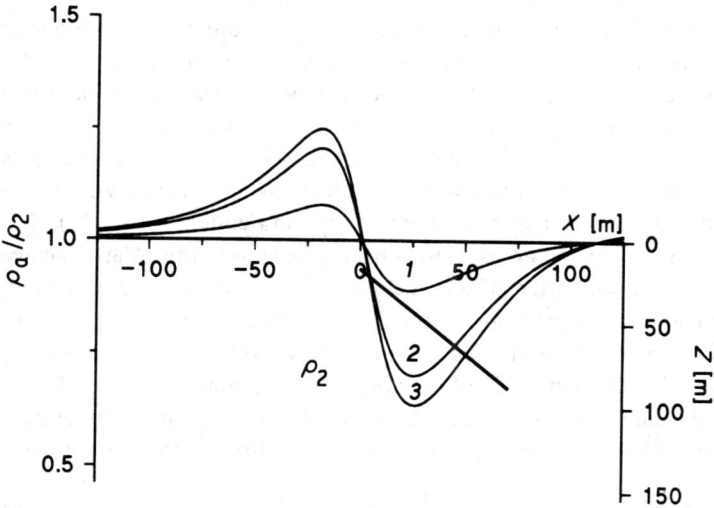

Figure 3.6 Apparent resistivity of a thin square conductor. Mid-gradient configuration, location of current electrodes $x = \pm 1000$ m, $p_2 = 1000\,\Omega$ m. Curve (1) $g = 0.05$ S; (2) $g = 0.5$ S; and (3) $g = 5$ S.

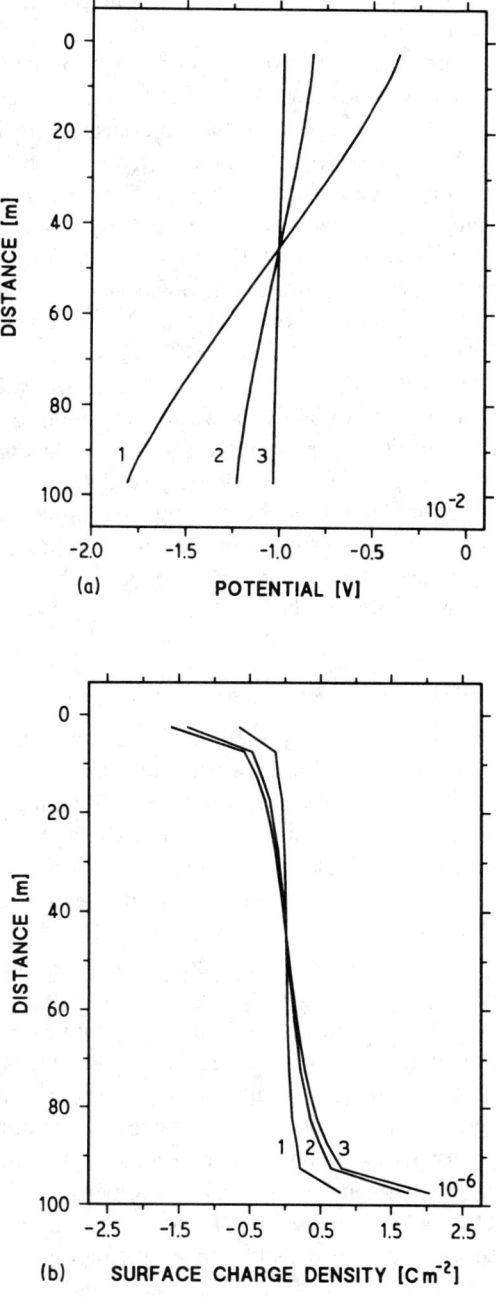

Figure 3.7 Variation of (a) potential and (b) charge density along the bisector of a conductor in the direction of dip, starting from the upper edge. Based on model presented in Fig. 3.6.

divided into 400 uniform square subsections and the potential of each subsection is assumed to be constant.

Fig. 3.6 depicts the apparent resistivity of the conductor on the ground surface. The dip of the conductor is apparent from the asymmetry of the anomaly curves and the anomaly amplitude increases with increasing longitudinal conductance. The development of the physical state of the conductor with increasing conductance can be examined from Figs 3.7(a) and 3.7(b), which illustrates the potential and the charge density along the conductor. At low conductance (curve 1), the potential is high on the side of positive current electrode and decreases steeply towards the negative one. At high conductance (curve 3), the potential is almost constant. The charge density increases with increasing conductance and the edge effect is visible as a steep rise in the charge density towards the edges. This is negative on the side of the positive current electrode but becomes positive towards the negative electrode in such a way that the total charge on the conductor is zero. At conductance $g = 5s$ (curve 3), the potential generated by the charge distribution is such that it makes the total potential of the conductor approximately constant. The dynamic counterpart of this charge distribution is the current flowing in the sheet with negligible resistance with the consequence that it short-circuits the potential differences in the vicinity of the conductor.

3.4 MAGNETOMETRIC RESISTIVITY

A special kind of resistivity survey method is the magnetometric resistivity (MMR) method [19], which is based on the measurement of the low-level magnetic field generated by a current introduced galvanically into the earth. The frequency of the current must be sufficiently low (1–5 Hz) so that the effect of electromagnetic induction is negligible. In comparison with more usual galvanic methods, the MMR method has the technical advantage that grounded electrodes must be used only for feeding the excitation current into the earth. Successful field tests have been made with this method [19], but so far it has not achieved particularly wide application.

In the earth model depicted in Fig. 3.8, an anomalous body with resistivity ρ_1 is located in a surrounding space with resistivity ρ_2. The system is excited by an electric current fed into the ground through grounded electrodes. Electric current passing through the anomalous body makes the surface of the body charged in conformity with the physical relations explained in Section 2.5. These surface charge distributions disturb the electric field and hence, in accordance with Ohm's law, the current flows in the body and in its environment. As a consequence of these disturbances in the currents, the magnetic field generated by the currents is also disturbed, and it is these magnetic anomalies which are of interest in MMR measurements.

From the foregoing it is clear that the calculation of MMR anomalies consists of two stages: determination of the current distributions that act as

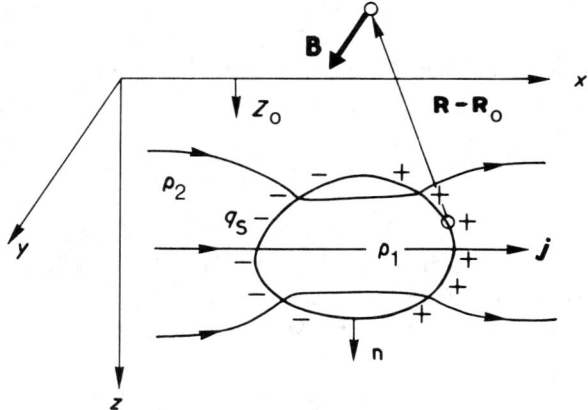

Figure 3.8 Physical model for the magnetometric resistivity method.

the effective sources of the magnetic field, and calculation of the magnetic field generated by these. To find out what information is needed concerning the current density, in order to be able to calculate the magnetic field, consider first, after Edwards *et al.* [19], the magnetic field generated by a laminar current flowing in the earth.

The solenoidal magnetic field $\mathbf{B}(\mathbf{R})$ due to a current flowing in volume V having density $\mathbf{j}(\mathbf{R}_0)$ can be expressed by Biot and Savart's law as follows:

$$\mathbf{B}(\mathbf{R}) = \mu \int_V \mathbf{j}(\mathbf{R}_0) \times \nabla_0 G_0(\mathbf{R}, \mathbf{R}_0) \, dV_0 \qquad (3.4.1)$$

where G_0 is the whole-space Green's function

$$G_0(\mathbf{R}, \mathbf{R}_0) = \frac{1}{4\pi |\mathbf{R} - \mathbf{R}_0|} \qquad (3.4.2)$$

V is the volume containing the current $\mathbf{j}$, μ is the free-space magnetic permeability and ∇_0 denotes differentiation with respect to the source point $\mathbf{R}_0$. Applying the partial differentiation rule, $\nabla \times (a\mathbf{A}) = a\nabla \times \mathbf{A} - \mathbf{A} \times \nabla a$, and the vector theorem

$$\int_V \nabla \times \mathbf{A} \, dV = \oint_S \mathbf{n} \times \mathbf{A} \, dS$$

where S is the surface bounding volume V and $\mathbf{n}$ is the unit vector along the outward normal of S, equation (3.4.1) becomes

$$\mathbf{B}(\mathbf{R}) = \mu \int_V G_0(\mathbf{R}, \mathbf{R}_0) \nabla_0 \times \mathbf{j}(\mathbf{R}_0) \, dV_0 - \mu \oint_S \mathbf{n} \times \{G_0(\mathbf{R}, \mathbf{R}_0) \mathbf{j}(\mathbf{R}_0)\} \, dS_0 \quad (3.4.3)$$

Consider the volume V to be the half-space $z > 0$, and the boundary S of V to consist of a plane just above the ground surface and a hemispherical surface

50 *Electrical methods*

of large radius r, which in the limit as r approaches infinity completely encloses V. The surface integral over S vanishes on the plane boundary where $j = 0$; on the large hemisphere it vanishes because j approaches zero faster than $1/r$, as r approaches infinity. Hence, equation (3.4.3) simplifies to

$$\mathbf{B}(\mathbf{R}) = \mu \int_V G_0(\mathbf{R}, \mathbf{R}_0)\, \nabla_0 \times \mathbf{j}\,(\mathbf{R}_0)\, dV_0 \qquad (3.4.4)$$

We next apply Ohm's law, $\mathbf{j} = g\mathbf{E}$ in equation (3.4.4), where $g = 1/\rho$ is the conductivity of the medium. Applying the partial differentiation rule, $\nabla \times \mathbf{j} = \nabla \times (g\mathbf{E}) = g\nabla \times \mathbf{E} - \mathbf{E} \times \nabla g$, and taking into consideration that, for static fields, $\nabla \times \mathbf{E} = 0$, equation (3.4.4) acquires the form

$$\mathbf{B}(\mathbf{R}) = \mu \int_V -G_0(\mathbf{R}, \mathbf{R}_0)\, \mathbf{E}(\mathbf{R}_0) \times \nabla_0 g(\mathbf{R}_0)\, dV_0 \qquad (3.4.5)$$

It is apparent from equation (3.4.5) that the only regions of V which contribute to the magnetic field are those in which the conductivity is varying. For the earth model under consideration, these are the surface of the anomalous body and the ground surface. The volume integral in equation (3.4.5) thus reduces to two surface integrals as follows:

$$\mathbf{B}(\mathbf{R}) = \mu \int_S (g_1 - g_2)\, G_0(\mathbf{R}, \mathbf{R}_0)\, \mathbf{E}(\mathbf{R}_0) \times \mathbf{n}\, dS_0$$

$$+ \mu \int_{z=0} -g_2\, G_0(\mathbf{R}, \mathbf{R}_0)\, \mathbf{E}(\mathbf{R}_0) \times \mathbf{z}_0\, dS_0 \qquad (3.4.6)$$

where S denotes the surface of the anomalous body and $z = 0$ is the ground surface. $\mathbf{n}$ is the outward unit normal vector on S and $\mathbf{z}_0$ is the unit vector in the direction of the z-axis.

The next step in the solution of the problem is to determine the tangential component of the electric field on surfaces S and $z = 0$. This can be done, for example, by using the integral equation technique considered in Section 3.3.2, by which the surface charge density q_S is first determined on boundary S of the anomalous body by integral equation (3.3.9):

$$\frac{g_1 + g_2}{2(g_1 - g_2)}\, q_S(\mathbf{R}) = -\frac{\partial U_P(\mathbf{R})}{\partial n} - \int_S \frac{\partial G(\mathbf{R}, \mathbf{R}_0)}{\partial n}\, q_S(\mathbf{R}_0)\, dS_0 \qquad (3.4.7)$$

The charge distribution on the ground surface is taken into consideration by applying the half-space Green's function G in determining q_S:

$$G(\mathbf{R}, \mathbf{R}_0) = \frac{1}{4\pi\,|\mathbf{R} - \mathbf{R}_0|} + \frac{1}{4\pi\,|\mathbf{R} - \mathbf{R}_0'|} \qquad (3.4.8)$$

where $\mathbf{R}_0'$ is the mirror image of $\mathbf{R}_0$ in relation to plane $z = 0$.

After q_S has been determined numerically from equation (3.4.7), the tangential component of the electric field can be calculated from the equation

$$\mathbf{n} \times \mathbf{E}(\mathbf{R}) = -\mathbf{n} \times \nabla U_P(\mathbf{R}) - \int_S \mathbf{n} \times \nabla G(\mathbf{R}, \mathbf{R}_0) \, q_S(\mathbf{R}_0) \, dS_0 \qquad (3.4.9)$$

where $\mathbf{R}_0$ is a point on boundary S, and $\mathbf{R}$ denotes the source points of equation (3.4.6) located on the surfaces S and $z = 0$. The tangential field component $\mathbf{n} \times \mathbf{E}$ calculated from equation (3.4.9) is substituted into equation (3.4.6), which can then be used for calculating the magnetic field at an arbitrary point in the space.

The magnetic field measured in the MMR survey, or obtained by model calculations, can be considered to be the sum of a primary field $\mathbf{B}_P$ and an anomalous field $\mathbf{B}_A$. The primary field is defined as the field which would be obtained if the anomalous body were not present. For the earth model composed of an anomalous body located in a homogeneous half-space, the primary magnetic field is thus the field generated by the primary current flow on the ground surface, which is deduced from equation (3.4.6) as

$$\mathbf{B}_P(\mathbf{R}) = \mu \int_{z=0} g_2 \, G_0(\mathbf{R}, \mathbf{R}_0) \, \nabla_0 U_P(\mathbf{R}_0) \times \mathbf{z}_0 \, dS_0 \qquad (3.4.10)$$

The x-, y- and z-components (Fig. 3.8) of the primary magnetic field on the ground surface due to a point electrode emitting current I at the origin, are obtained from equation (3.4.10) in the following form [19]:

$$B_{Px}(x, y, 0) = -\frac{\mu I}{4\pi} \frac{y}{x_2 + y_2}$$

$$B_{Py}(x, y, 0) = \frac{\mu I}{4\pi} \frac{x}{x_2 + y_2}$$

$$B_{Pz} = 0$$

The MMR anomaly is defined basically as an anomalous field component B_{Aa} in the direction a, expressed as a percentage of a definite primary magnetic field, for example the horizontal component B_{Pc} at point $(x', y', 0)$ in direction c:

MMR anomaly $(x, y, 0) = 100 \, B_{Aa}(x, y, 0) / B_{Pc}(x', y', 0)$

where the anomalous field is obtained by subtracting the primary field component from the corresponding measured field component:

$$B_{Aa}(x, y, 0) = B_a(x, y, 0) - B_{Pa}(x, y, 0)$$

Theoretical MMR anomalies have been mainly reported for models for which the electrical potential and thus the current density can be determined analytically.

3.5 *MISE-À-LA-MASSE* METHOD

In the *mise-à-la-masse* method, the conductive body to be surveyed is excited by a galvanic current fed directly into the body. The other current electrode is earthed at a sufficient distance that its effect on the potentials measured in the vicinity of the body is negligible. The excited conductor behaves like a large electrode strongly influencing the potential distribution due to the current.

The *mise-à-la-masse* method is best suited to delineating highly conductive bodies, but has also been applied to weakly conductive formations, such as fracture zones.

From the physical viewpoint, the *mise-à-la-masse* method can be considered to be a special case of the resistivity method, although it differs from conventional methods in its mathematical features. The primary field of the *mise-à-la-masse* model becomes small as the conductivity contrast between the body and its environment increases, which makes the integral equation representing the *mise-à-la-masse* system homogeneous. Considering that the integral equation for the resistivity model is non-homogeneous, it is clear that the solution properties of these systems differ from each other.

In the following we shall consider the application of three-dimensional integral equations of charge density and potential for calculating *mise-à-la-masse* anomalies.

3.5.1 Integral equations for charge density

The earth model for the *mise-à-la-masse* system is illustrated in Fig. 3.9. A current I is fed into the conductor having resistivity ρ_1. The resistivity of the surrounding medium is $\rho_2 > \rho_1$. The current leaving the body generates a positive charge distribution of density $q_S\,(=\sigma_S/\varepsilon_0)$ on the surface S of the body. This charge is the source of the *mise-à-la-masse* anomaly.

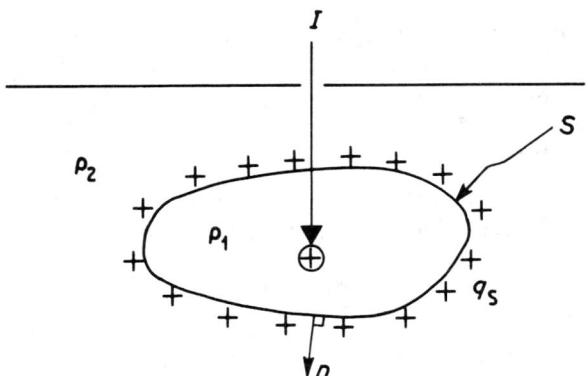

Figure 3.9 The *mise-à-la-masse* model.

The potential generated by the system can be written in accordance with equation (3.3.8) in Section 3.3.2 as follows:

$$U(\mathbf{R}) = U_P(\mathbf{R}) + \int_S G(\mathbf{R}, \mathbf{R}_0) \, q_S(\mathbf{R}_0) \, dS_0 \qquad (3.5.1)$$

where S denotes the total surface of the anomalous region, which may be composed of one body or several separate bodies. The primary potential U_P due to a point electrode earthed in the body is

$$U_P(\mathbf{R}) = \rho_1 G(\mathbf{R}, \mathbf{R}_P) \, I(\mathbf{R}_P) \qquad (3.5.2)$$

G is Green's function for the space surrounding the body. Note that the resistivity appearing in equation (3.5.2) is that of the medium containing the current earthing, that is, in the *mise-à-la-masse* system, the resistivity of the conductor.

To determine the unknown surface charge density q_S, we form an integral equation by differentiating both sides of equation (3.5.1) and moving the calculation point $\mathbf{R}$ to the surface S. The singularity of Green's function at $\mathbf{R} = \mathbf{R}_0$ is treated similarly as in the derivation of equation (3.3.9). Substituting q_S for the normal derivative of U according to equation (3.3.2), we obtain the non-homogeneous Fredholm integral equation of the second kind:

$$\frac{\rho_2 + \rho_1}{2(\rho_2 - \rho_1)} q_S(\mathbf{R}) = -\frac{\partial U_P(\mathbf{R})}{\partial n} - \int_S \frac{\partial G(\mathbf{R}, \mathbf{R}_0)}{\partial n} q_S(\mathbf{R}_0) \, dS_0 \qquad (3.5.3)$$

Equation (3.5.3) has a unique solution as long as the primary field (the first term on the right-hand side of the equation) is of considerable strength in relation to the secondary field generated by the surface integral. Model calculations have indicated that this condition is satisfied so long as $\rho_1 > 0.1 \rho_2$. As ρ_1 goes below this limit, the anomaly intensities calculated from equation (3.5.3) start to collapse. This is because the primary field obtained from equation (3.5.2) approaches zero with decreasing ρ_1 and thus equation (3.5.3) becomes a homogeneous integral equation. Fredholm's theorem states that, if a non-homogeneous integral equation of the second kind has a unique solution, then the corresponding homogeneous equation does not have solutions. Consequently, for highly conductive bodies, the *mise-à-la-masse* problem cannot be directly solved from equation (3.5.3).

In the following we shall consider two methods of proceeding with the problem. First, equation (3.5.3) can be substituted by the corresponding equation written for the reciprocal electrode configuration [20], which means that the current electrode is considered to be earthed at the calculation point and the potential generated by this electrode is then calculated at the true earthing point in the body. By this method, the resistivity appearing in the primary potential is ρ_2, and the integral equation (3.5.3) thus obtained is non-homogeneous and can be solved directly by standard numerical methods.

A disadvantage in solving the *mise-à-la-masse* model from the integral equation associated with the reciprocal electrode configuration, is the slow convergence of solutions when the observation point is near the body. Furthermore, the principle of reciprocity is of no use when the observation point is in the body housing the current electrode.

An elegant method for solving q_S from equation (3.5.3) with the true electrode configuration is the introduction of a continuity equation into the system [21]. The application of the continuity equation together with equation (3.5.3) results in a set of equations which is overdetermined, that is, the number of equations exceeds the number of unknown q_S values. Therefore, the set is solved in the least-squares sense. Provided the continuity equations for the surface $S(I)$ enclosing the current electrode

$$\oint_{S(I)} q_S \, dS = \rho_2 I \tag{3.5.4}$$

and for other surfaces $S(\text{no } I)$

$$\oint_{S(\text{no } I)} q_S \, dS = 0 \tag{3.5.5}$$

are satisfied exactly, equation (3.5.3) is satisfied in the least-squares sense.

The solutions obtained with the least-squares procedure are accurate even in cases of very high resistivity contrasts [21]. The high accuracy of the anomaly intensities is due to the application of continuity equations, which constrain the total amount of charge to be correct on each boundary surface.

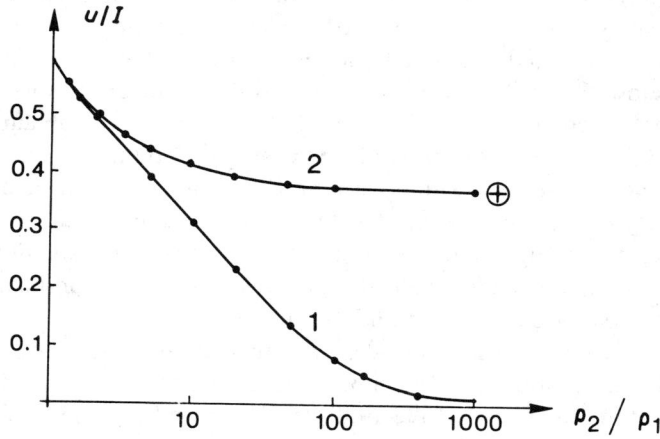

Figure 3.10 *Mise-à-la-masse* anomaly on the ground surface above the centre of a cube calculated from the integral equation of surface charge (1) and with the least-squares solution (2). $\rho_2 = 100 \, \Omega\text{m}$. The anomaly calculated using the method based on equipotentiality is marked by a cross (+). After [21].

Figure 3.10 illustrates the *mise-à-la-masse* anomaly above a cube, calculated from equation (3.5.3), and from equations (3.5.3)–(3.5.5) using the least-squares procedure [21]. As the resistivity contrast increases, the anomaly calculated solely from equation (3.5.3) becomes inaccurate, whereas the anomaly calculated with the least-squares procedure behaves correctly, approaching the anomaly value (marked by a cross (+) in the figure) calculated by assuming that the conductor is at constant potential. This calculation method shall be considered in Section 3.5.3.

3.5.2 Integral equation for potential

In Section 3.3.3 we formulated the solution to resistivity problems with the aid of an integral equation for potential. The source of the secondary potential was represented by a non-physical double layer distributed over the boundary of the anomalous body. We shall now follow a similar technique in formulating the solution of the *mise-à-la-masse* problem.

The earth model to be considered is illustrated in Fig. 2.1. S is the surface of the conducting body V_1 and A denotes the boundaries of space V_2 surrounding the body. Surface B symbolizes the limit of the space at infinity. The resistivities ρ_1 and ρ_2 of regions V_1 and V_2 are homogeneous. The primary current electrode is earthed in the conductor V_1.

The potential U due to current I emitted by a point electrode into the conductor satisfies Poisson's equation in region V_1:

$$\nabla^2 U = -\rho_1 I \delta(\mathbf{R} - \mathbf{R}_P) \tag{3.5.6}$$

where δ is Dirac's delta function and $\mathbf{R}_P$ the earthing point of the current electrode. In regions V_2 and V_3, U satisfies Laplace's equation:

$$\nabla^2 U = 0 \tag{3.5.7}$$

The solutions of equations (3.5.6) and (3.5.7) must satisfy the boundary conditions:

$$U_1 = U_2$$

$$\frac{1}{\rho_1}\frac{\partial U_1}{\partial n} = \frac{1}{\rho_2}\frac{\partial U_2}{\partial n} \quad \text{on} \quad S \tag{3.5.8}$$

and

$$U_2 = U_3$$

$$\frac{1}{\rho_2}\frac{\partial U_2}{\partial n} = \frac{1}{\rho_3}\frac{\partial U_3}{\partial n} \quad \text{on} \quad A \tag{3.5.9}$$

To convert the boundary-value problem defined by equations (3.5.6)–(3.5.9) into an integral equation, use is made of Green's function G which is to satisfy Poisson's equation

56   Electrical methods

$$\nabla^2 G(\mathbf{R}, \mathbf{R}_0) = -\delta(\mathbf{R} - \mathbf{R}_0) \tag{3.5.10}$$

Like the potential U, the solution of equation (3.5.10) must satisfy the boundary conditions

$$G_2 = G_3$$
$$\frac{1}{\rho_2}\frac{\partial G_2}{\partial n} = \frac{1}{\rho_3}\frac{\partial G_3}{\partial n} \quad \text{on} \quad A \tag{3.5.11}$$

On surface S, no boundary conditions are imposed upon G.

Functions U and G are next substituted into Green's second identity. Applying the identity first to region V_2 so that the operations are performed with respect to variable $\mathbf{R}$, and by keeping point $\mathbf{R}_0$ in V_2, we obtain:

$$\int_{V_2} (G_2 \nabla^2 U_2 - U_2 \nabla^2 G_2)\, dV = \int_S \left(-G_2 \frac{\partial U_2}{\partial n} + U_2 \frac{\partial G_2}{\partial n}\right) dS$$

$$+ \int_A \left(G_2 \frac{\partial U_2}{\partial n} - U_2 \frac{\partial G_2}{\partial n}\right) dA \tag{3.5.12}$$

Applying Green's second identity to region V_3 and taking into consideration that the regularity properties of U and G make the integral over surface B vanish, we further obtain

$$0 = \int_A \left(-G_3 \frac{\partial U_3}{\partial n} + U_3 \frac{\partial G_3}{\partial n}\right) dA \tag{3.5.13}$$

Multiplying equation (3.5.13) by ρ_2/ρ_3, adding the respective sides of equations (3.5.13) and (3.5.12) and taking into consideration the boundary conditions (3.5.9) and (3.5.11), it follows that the surface integral in equation (3.5.12) over A vanishes. Substituting from equations (3.5.7) and (3.5.10) and dropping the subscript from G (G is continuous when point $\mathbf{R}$ is moved through surface S keeping point $\mathbf{R}_0$ fixed), equation (3.5.12) becomes

$$U_2(\mathbf{R}_0) = \int_S \left[-G(\mathbf{R}, \mathbf{R}_0) \frac{\partial U_2(\mathbf{R})}{\partial n} + U_2(\mathbf{R}) \frac{\partial G(\mathbf{R}, \mathbf{R}_0)}{\partial n}\right] dS \tag{3.5.14}$$

Applying Green's second identity to region V_1 keeping point $\mathbf{R}_0$ in region V_2 and substituting equations (3.5.6) and (3.5.10), equation (3.5.12) yields:

$$-\rho_1 G(\mathbf{R}, \mathbf{R}_P) I(\mathbf{R}_P) = \int_S \left[G(\mathbf{R}, \mathbf{R}_0) \frac{\partial U_1(\mathbf{R})}{\partial n} - U_1(\mathbf{R}) \frac{\partial G(\mathbf{R}, \mathbf{R}_0)}{\partial n}\right] dS \tag{3.5.15}$$

Next we multiply equation (3.5.15) by ρ_2/ρ_1 and add this equation to equation (3.5.14). Substituting the boundary conditions (3.5.8) into the equation thus obtained, replacing $\mathbf{R}_0$ by $\mathbf{R}$, and applying the reciprocity therem, $G(\mathbf{R}_0, \mathbf{R}) = G(\mathbf{R}, \mathbf{R}_0)$, we obtain for potential U in region V_2:

$$U(\mathbf{R}) = \frac{\rho_2}{\rho_1} U_P(\mathbf{R}) + \frac{\rho_1 - \rho_2}{\rho_1} \int_S G'(\mathbf{R}, \mathbf{R}_0) U(\mathbf{R}_0) \, dS_0, \quad \mathbf{R} \text{ in } V_2 \quad (3.5.16)$$

where $G' = \partial G / \partial n_0$ and, in accordance with equation (2.3.5),

$$U_P(\mathbf{R}) = \rho_1 G(\mathbf{R}, \mathbf{R}_P) I(\mathbf{R}_P) \quad (3.5.17)$$

To derive the representation for U in region V_1, point $\mathbf{R}_0$ is located in V_1 when Green's second identity is applied to regions V_1 and V_2. Following a procedure similar to that above, we obtain for U in V_1:

$$U(\mathbf{R}) = U_P(\mathbf{R}) + \frac{\rho_1 - \rho_2}{\rho_2} \int_S G'(\mathbf{R}, \mathbf{R}_0) U(\mathbf{R}_0) \, dS_0, \quad \mathbf{R} \text{ in } V_1 \quad (3.5.18)$$

where U_P is as given in equation (3.5.17).

To find the potential U on surface S, an integral equation is developed from either equation (3.5.16) or equation (3.5.18) by moving the calculation point $\mathbf{R}$ to the surface S. Removing the effect of the singular point $\mathbf{R} = \mathbf{R}_0$ from the surface integral in accordance with equation (2.6.4), a Fredholm integral equation of the second kind is obtained for U:

$$\frac{\rho_1 + \rho_2}{2\rho_1} U(\mathbf{R}) = \frac{\rho_2}{\rho_1} U_P(\mathbf{R}) + \frac{\rho_1 - \rho_2}{\rho_1} \int_S G'(\mathbf{R}, \mathbf{R}_0) U(\mathbf{R}_0) \, dS_0 \quad (3.5.19)$$

where the surface integral is understood as the principal value. Equation (3.5.19) is solved for U applying numerical methods. Substituting the values of U thus obtained, the potential can be calculated in region V_2 from equation (3.5.16) and in region V_1 from equation (3.5.18).

Note that, because of the coefficient ρ_2/ρ_1 in front of the primary potential term U_P, the integral equation (3.5.19) is completely analogous with equation (3.3.26) and thus remains non-homogeneous regardless of the value of ρ_1. Therefore, its range of validity can extend to the resistivity contrast $\rho_2/\rho_1 = 10^4$, as demonstrated by Eloranta [13]. At this contrast, the model has achieved electrostatic saturation, which means that the calculations can also be made by assuming that the potential is constant in the conductor.

3.5.3 Integral equation for perfect conductor model

When the resistivity of an anomalous body decreases in relation to that of the environment, the charge distributed on its surface S becomes saturated. In the saturated state the potential of the body is constant. This equipotential condition can be used to form a simple and efficient integral equation solution for the potential problem for a perfect conductor model.

Consider the perfect conductor model for which, in Fig. 3.9, resistivity $\rho_1 = 0$. A current I fed into the conductor generates a positive charge distribution of density q_S on the surface S. Taking into consideration that, for the case $\rho_1 = 0$, the primary potential U_P represented by equation (3.5.2) is 0, the potential U due to the model can be represented in the form

58   *Electrical methods*

$$U(\mathbf{R}) = \int_S G(\mathbf{R}, \mathbf{R}_0)\, q_S(\mathbf{R}_0)\, \mathrm{d}S_0 \qquad (3.5.20)$$

where charge density q_S is unknown.

To obtain q_S, an integral equation is formed from equation (3.5.20) by moving the calculation point $\mathbf{R}$ to the surface S and regarding the potential on S as constant ($U = U_C$). As the singular point $\mathbf{R} = \mathbf{R}_0$ does not generate any additional term to U, we obtain for U on S:

$$U_C = \int_S G(\mathbf{R}, \mathbf{R}_0)\, q_S(\mathbf{R}_0)\, \mathrm{d}S_0, \quad \text{on } S \qquad (3.5.21)$$

If the anomalous region is composed of several separate bodies, S denotes the total surface of all the bodies, whereas on the left-hand side of equation (3.5.21) the constant potential U_C takes a specific value for each separate body.

As both q_S and U_C are unknown, equation (3.5.21) is underdetermined, and to solve the problem one additional equation is needed for each U_C value. These additional equations may again take the form of the continuity equation

$$\oint_{S(I)} q_S\, \mathrm{d}S = \rho_2 I \qquad (3.5.22)$$

for the surface $S(I)$ enclosing the current earthing, and for surfaces $S(\text{no } I)$ not enclosing any current earthings

$$\oint_{S(\text{no } I)} q_S\, \mathrm{d}S = 0 \qquad (3.5.23)$$

When S is composed of several separate surfaces, equations (3.5.22) and (3.5.23) are applied separately to each of them.

The unknown values of surface charge density q_S, together with the constant potential U_C are solved from the set of algebraic equations obtained by discretizing equations (3.5.21)–(3.5.23). After substituting q_S, U can be calculated at an arbitrary point in space from the discretized form of equation (3.5.20).

Consider a numerical example as an application of the simple calculation method given above. Figure 3.11 represents the *mise-à-la-masse* anomaly of an isolated conductor situated in a half-space, and with a screening conductive plate above it [22]. The half-space is supposed to have two parts separated by a vertical contact. The conductor and the plate are truncated prismatic bodies, which means that the contributions of their end faces are disregarded in the calculations. The solution is based on equipotentiality conditions for the conductor and the plate. Green's function for the half-space containing a vertical contact is given in Appendix B.

The current channeling in the screening plate is apparent as the reduced intensity and smooth shape of the anomaly curve (2). The contribution of the negative surface charge distributed on the vertical contact can be seen as the drop in intensity on the flanks of the anomaly above the contact.

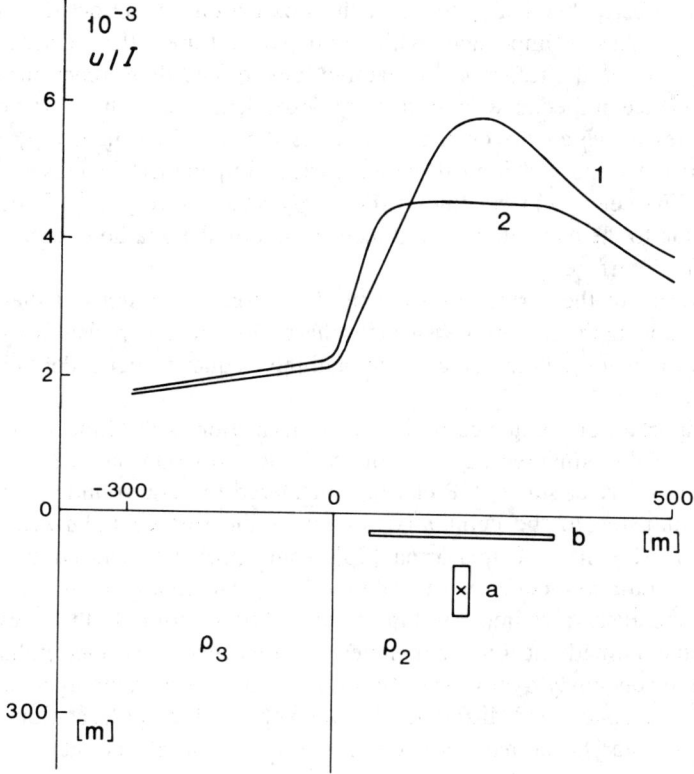

Figure 3.11 *Mise-à-la-masse* anomaly for a conductive truncated cylindrical body, located in a half-space with a vertical contact, alone (body a, curve 1), and with a shading conductive plate above it (bodies a and b, curve 2). $\rho_2 = 100\,\Omega\text{m}$ and $\rho_3 = 10\,\Omega\text{m}$. Length of the body 1000 m. After [22].

3.6 SURFACE POLARIZATION

3.6.1 Introduction

So far we have considered only electrical models in which the current carriers can freely penetrate the resistivity boundaries. The physical sources of the secondary electric field generated by the current flowing through these contacts are simple surface charge distributions.

In this section we shall deal with an electric model containing surface polarizable interfaces [23]. Surface polarization is generated on the boundaries between media having physically different current carriers. A typical surface polarization model encountered in electrical surveying is an electronic conductor, such as a metal or sulphide body, located in an electrolyte environment. Charge transfer across the metal–electrolyte interface can take place

60   *Electrical methods*

only if an electrochemical reaction such as oxidation–reduction occurs. Thus, the interface has an impedance which is dependent upon the electrochemical reactions and on the transport of reactive ions to feed these reactions.

The surface impedance is frequency-dependent and is high at very low frequencies at which the current penetrates the interface mainly by electrochemical reactions and ion diffusion. Surface impedance decreases with increasing frequency and becomes vanishingly small at very high frequencies. This is due to the capacitive admittance caused by the charge system oscillating on the interface.

As a result of the surface impedance, the secondary sources of the electric field include both the simple surface charge due to the resistivity contrast between the metal and the electrolyte, and the frequency-dependent electrical double layer.

An important consequence of surface polarization is the induced polarization (IP) of disseminated sulphide or oxide deposits. On the scale of a conventional electrical survey, IP can be considered to be a volume property of the medium due to the cumulative effect of the surface polarized mineral grains. As shown by Guptasarma [23], some problems caused by surface polarization are associated with scale model measurements of resistivity and IP using metal samples immersed in an electrolyte solution. In these resistivity measurements made at low frequencies, the presence of surface polarization has been erroneously overlooked. In the IP model measurements using metal samples, the conceptual difference between the surface polarization and the IP volume polarization has not been properly taken into account.

3.6.2 Solution for perfect conductor models with surface polarization

In the surface polarization model, the secondary potential is generated by the simple surface charge σ_S and the electric double layer τ, provoked by the electric current flowing across the interface S between the electronic conductor and the electrolyte (see Fig. 3.12). The potential-theory properties of simple and double sources were considered in Sections 2.5 and 2.6, in accordance with which the potential of the surface polarization model can be written in the form:

$$U(\mathbf{R}) = U_P(\mathbf{R}) + \frac{1}{\varepsilon_0} \int_S \{G(\mathbf{R}, \mathbf{R}_0) \, \sigma_S(\mathbf{R}_0) + G'(\mathbf{R}, \mathbf{R}_0) \, \tau(\mathbf{R}_0)\} \, \mathrm{d}S_0 \quad (3.6.1)$$

where U_P is the primary potential and $G' = \partial G/\partial n_0$. For the half-space environment, Green's function for the simple source is

$$G(\mathbf{R}, \mathbf{R}_0) = \frac{1}{4\pi |\mathbf{R} - \mathbf{R}_0|} + \frac{1}{4\pi |\mathbf{R} - \mathbf{R}_0'|} \quad (3.6.2)$$

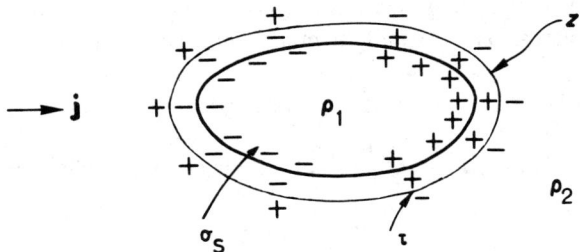

Figure 3.12 Surface polarization model.

where $\mathbf{R}_0'$ is the image of point $\mathbf{R}_0$ in relation to the surface of the half-space.

The surface charge density σ_S and the electrical double layer τ in equation (3.6.1) are unknown functions. Our aim is now to represent σ_S and τ in terms of a proper field vector common to both these functions. It is apparent from equations (2.5.7) and (2.6.6), that σ_S and τ can both be expressed in terms of the normal component of current density j_n as follows:

$$\sigma_S(\mathbf{R}_0) = \varepsilon_0 (\rho_2 - \rho_1) j_n(\mathbf{R}_0) \qquad (3.6.3)$$

$$\tau(\mathbf{R}_0) = -\varepsilon_0 Z j_n(\mathbf{R}_0) \qquad (3.6.4)$$

Substituting equations (3.6.3) and (3.6.4) into equation (3.6.1) we get potential U as a function of the unknown j_n:

$$U(\mathbf{R}) = U_P(\mathbf{R}) + \int_S \left[(\rho_2 - \rho_1) G(\mathbf{R}, \mathbf{R}_0) - Z G'(\mathbf{R}, \mathbf{R}_0) \right] j_n(\mathbf{R}_0) \, dS_0 \qquad (3.6.5)$$

To solve for j_n, we deduce an integral equation from equation (3.6.5) by moving the point $\mathbf{R}$ to the surface S on the inner side V_1 of S. When the singular point $\mathbf{R} = \mathbf{R}_0$ is removed from the integral, the double layer term introduces an additional term in the integral equation according to equation (2.6.4b). Taking into account that the resistivity of the electronic conductor is generally several orders of magnitude lower than that of the electrolyte, $\rho_1 \ll \rho_2$, we can assume that the region V_1 inside S is at constant potential $U_1 = U_C$. Substituting equation (3.6.4) into the term due to the singularity we now get the integral equation for j_n:

$$U_C = U_P(\mathbf{R}) + \frac{Z}{2} j_n(\mathbf{R})$$
$$+ \int_S \left[\rho_2 G(\mathbf{R}, \mathbf{R}_0) - Z G'(\mathbf{R}, \mathbf{R}_0) \right] j_n(\mathbf{R}_0) \, dS_0 \qquad (3.6.6)$$

where the surface integral is to be considered as a principal value.

62   *Electrical methods*

Since both U_C and j_n are unknown, equation (3.6.6) is underdetermined and one additional equation is needed to solve the problem. We again use the continuity equation as the supplementary equation:

$$\int_{S(no\,I)} j_n \, dS = 0 \qquad (3.6.7)$$

for surfaces $S(no\,I)$ which do not enclose current electrodes, and

$$\int_{S(I)} j_n \, dS = I \qquad (3.6.8)$$

for surfaces $S(I)$ which enclose an electrode emitting current I.

Equations (3.6.6)–(3.6.8) are discretized and the set of linear algebraic equations thus obtained is solved numerically for j_n and U_C. After substitution of j_n, the potential U can be calculated in the environment V_2 of the conductor from equation (3.6.5).

If the model is composed of several separate conducting bodies, equations (3.6.7) and (3.6.8) are applied separately to each conductor. Each conductor, moreover, achieves its own U_C whereas the surface integral in equation (3.6.6) is extended over the total surface of all the conductors. The impedance Z is frequency-dependent, but it may also be a function of location $\mathbf{R}_0$.

To demonstrate the effect of surface polarization on the physical scale modelling results, we calculate, using the above formulation, the apparent

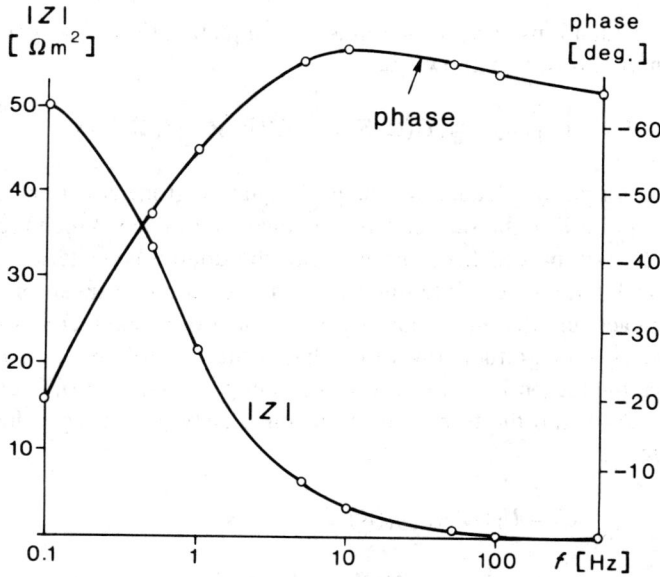

Figure 3.13 Magnitude and phase spectra of surface impedance between aluminium and water. After [24].

resistivity anomalies of an aluminum plate (of dimensions 0.01 m × 0.1 m × 0.5 m) immersed in tap water. The calculations are made for surface impedance values $Z(f)$ measured in the laboratory at frequencies, f, of 0.1 Hz, 0.5 Hz, 1 Hz, 5 Hz, 10 Hz, 50 Hz, 100 Hz and 500 Hz [24]. The magnitude and phase of the complex impedance obtained are shown in Fig. 3.13. The electrode array used is the Wenner array with $a = 0.02$ m. The apparent resistivity $\rho_a(f)$ is calculated from the magnitudes of the complex voltage values $\Delta U(f)$ between the potential electrodes from the formula

$$\rho_a(f) = 2\pi a |\Delta U(f)|/I \qquad (3.6.9)$$

The complex potential $U(f)$ is calculated numerically from equations (3.6.6) and (3.6.5) by using pulse functions as a subsectional basis for current density j_n.

The calculated apparent resistivity anomalies are shown in Fig. 3.14. The ρ_a anomaly corresponding to frequency $f = 500$ Hz, at which the magnitude of the surface impedance is low (see Fig. 3.13), is similar to a conventional resistivity anomaly for a highly conductive plate. As $|Z|$ is low, the density

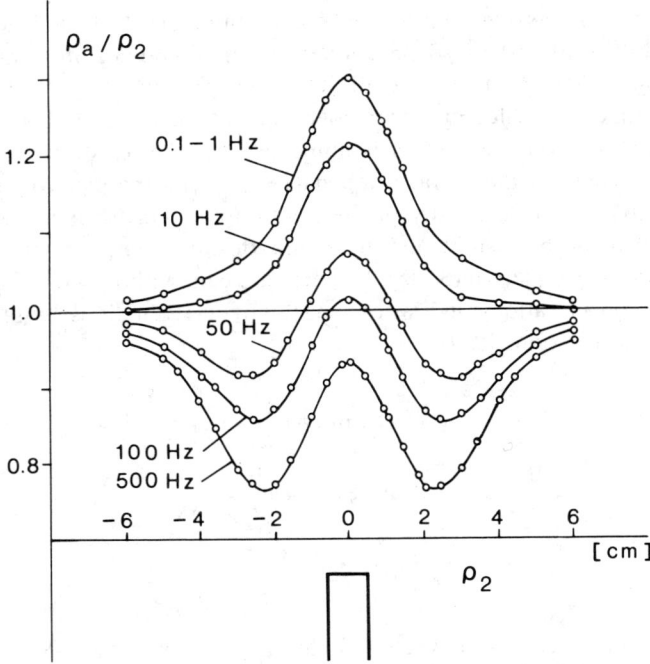

Figure 3.14 Calculated apparent resistivity of an aluminium sheet immersed in tap water. Resistivity of water $\rho_2 = 69.8$ Ω m and surface impedance values are from Fig. 3.13. After [24].

of the electric double layer on surface S is also low and the ρ_a anomaly is thus generated by the simple surface charge distribution. With decreasing frequency the magnitude of the surface impedance increases, which makes the density of the double layer increase and the density of the simple surface charge decrease. At $f = 1\,\text{Hz}$, Z has achieved such a high value that the double layer is saturated and the simple surface charge has vanished. As a result, the resistivity anomaly of an insulator is obtained! However, this anomaly is not similar to that which would be obtained if the perfect conductor with high surface impedance were replaced by a thoroughly resistive body, for which the secondary sources are simple surface charge distributions.

It is worth nothing that the model considered above is comparable with a physical scale modelling case given by Guptasarma [23]. The calculated and measured anomalies agree well, taking into consideration that the Z and ρ_2 values used in calculations are obviously not identical with the values that prevailed in the measurements reported by Guptasarma [23].

3.6.3 Scaling of surface polarization models

The dimensions of surface polarizable bodies contained within geological formations vary in size from those of an individual sulphide or oxide mineral grain to those of massive occurrences of these minerals. To be able to estimate the contribution of surface polarization occurring on bodies of different sizes, it is of importance to know the scaling relations for surface polarization.

Consider a current I fed by a point electrode into an electrolyte environment in which an electronic conductor is immersed. Let the resistivity of the electrolyte be ρ and let the surface impedance between the electrolyte and the conductor be Z. The resistivity of the electronic conductor is assumed to be so low that it can be disregarded in calculations. The potential of this model is represented by equation (3.6.5), in which we can, without loss of generality in scaling considerations, use whole-space Green's functions by making the following substitutions [25]:

$$G(\mathbf{R}, \mathbf{R}_0) = \frac{1}{4\pi r}$$

$$G(\mathbf{R}, \mathbf{R}_P) = \frac{1}{4\pi r_P}$$

$$G'(\mathbf{R}, \mathbf{R}_0) = \frac{\mathbf{r} \cdot \mathbf{n}}{4\pi r^3}$$

where $\mathbf{n}$ is the normal unit vector of the polarizable surface S. With these substitutions we obtain from equation (3.6.5) the following expression for U:

$$U = \frac{\rho I}{4\pi r_P} + \frac{\rho}{4\pi} \int_S \left(\frac{1}{r} - \frac{Z}{\rho} \frac{\mathbf{r} \cdot \mathbf{n}}{r^3} \right) j_n \, dS_0 \qquad (3.6.10)$$

We examine next two models that are identical in terms of geometry but differ in scale. Let the characteristic dimension of the smaller model S be a_S and that of the larger model L, a_L. We define the scaling factor k between models S and L to be

$$k = a_L/a_S \, (> 1) \qquad (3.6.11)$$

The apparent resistivity ρ_a of the model is proportional to the product of potential U and the geometric factor of the electrode array. The geometric factor is proportional to the dimension of the electrode array, and hence to the scaling factor k. In accordance with equation (3.6.10), the apparent resistivities ρ_{aS} and ρ_{aL} of the small and large models are as follows:

$$\rho_{aS} \propto \frac{\rho_S I}{4\pi r_{PS}} + \frac{\rho_S}{4\pi} \int_{S_S} \left(\frac{1}{r_S} - \frac{Z_S}{\rho_S} \frac{\mathbf{r}_S \cdot \mathbf{n}}{r_S^3} \right) j_n \, dS_{0S} \qquad (3.6.12a)$$

and

$$\rho_{aL} \propto \frac{k\rho_L I}{4\pi r_{PL}} + \frac{\rho_L}{4\pi} \int_{S_L} \left(\frac{k}{r_L} - \frac{kZ_L}{\rho_L} \frac{\mathbf{r}_L \cdot \mathbf{n}}{r_L^3} \right) j_n \, dS_{0L} \qquad (3.6.12b)$$

The larger model can be expressed in terms of the dimensions of the smaller model by substituting $r_L = k r_S$, hence:

$$\rho_{aL} \propto \frac{\rho_L I}{4\pi r_{PS}} + \frac{\rho_L}{4\pi} \int_{S_S} \left(\frac{1}{r_S} - \frac{Z_L}{k\rho_L} \frac{\mathbf{r}_S \cdot \mathbf{n}}{r_S^3} \right) j_n \, dS_{0S} \qquad (3.6.12c)$$

It is now possible to examine the relation of the smaller model to the equivalent larger model by comparing equation (3.6.12a) and (3.6.12c). In requiring complete equivalence of the models, it is obvious that the geometry of the current density pattern j_n must also be preserved when the scale changes. By requiring that the relative apparent resistivities of the models are equal:

$$(\rho_a/\rho)_S = (\rho_a/\rho)_L \qquad (3.6.13)$$

the following scaling relation for the physical quantities Z and ρ is obtained:

$$\frac{Z_S}{\rho_S} = \frac{Z_L}{k\rho_L} \qquad (3.6.14)$$

The physical meaning of the scaling relation (3.6.14) can be illustrated by considering a model in which surface impedance Z is formed of a material layer with resistivity ρ_Z and thickness t, such that $Z = \rho_Z t$. If this model is scaled by factor k while keeping the resistivity values unchanged, then the thickness of the resistive layer becomes kt and, in accordance with equation (3.6.14), the surface impedance is thus scaled to $Z = k\rho_Z t$.

The above scaling considerations make it clear that the effect of surface polarization becomes stronger as the polarized bodies which generate the effect become smaller. For example, it readily explains the generation of IP volume polarization in a disseminated ore deposit by a large amount of small sulphide mineral grains. On the other hand, the effect of surface impedance on the interface of a large massive sulphide deposit is of minor importance in relation to the strong current conduction in the deposit. It is also important to note that large errors may occur in physical scale modelling using electronic conductors, if the effect of surface impedance is not properly compensated for in the modelling arrangement.

3.7 INDUCED POLARIZATION

3.7.1 Introduction

It was explained in Section 3.2 that the resistivity of rocks containing disseminated surface-polarizable minerals is complex and frequency-dependent – a property generally termed **resistivity dispersion**. This phenomenon is also caused by the presence of ion-exchanging minerals in the rocks. The microscopic processes that are responsible for resistivity dispersion are considered in more detail by Sumner [26].

Resistivity dispersion of rocks is the basis for the induced polarization (IP) method. The shape of the dispersion curve is characteristic of the texture of mineralization and thus also of the geological history of the formation to be surveyed.

In the following we shall first consider some theoretical relations for linear systems which are important for understanding the basic principles concerning the frequency and time dependence of resistivity. Other important subjects that we consider are the potential-theory methods for calculating IP anomalies in space coordinates, using linear integral equations.

3.7.2 Information-theoretical relations concerning resistivity dispersion

As stated in Section 3.2, the IP method is applied at such a low current density that the system is linear. Consequently, the information theoretical properties of the IP system can be treated by means of linear systems theory [27].

Consider the IP system on the basis of Ohm's law, which is represented in the frequency domain as

$$E(\omega) = \rho(\omega) j(\omega) \qquad (3.7.1)$$

where $\omega = 2\pi f$ is the angular frequency of current, E is the electric field strength, j the current density, and ρ is the resistivity of the medium.

To transform this equation into the time domain we define the following Fourier transformation pair:

$$F\{f(t)\} = \bar{f}(\omega) = \int_{-\infty}^{\infty} f(t) e^{-i\omega t} \, dt \qquad (3.7.2a)$$

$$F^{-1}\{\bar{f}(\omega)\} = f(t) = \frac{1}{2\pi} \int_{-\infty}^{\infty} \bar{f}(\omega) e^{i\omega t} \, d\omega \qquad (3.7.2b)$$

where $i^2 = -1$.

Denoting $E(t) = F^{-1}\{E(\omega)\}$, $j(t) = F^{-1}\{j(\omega)\}$, and $R(t) = F^{-1}\{p(\omega)\}$, we obtain the time-domain Ohm's law in the form of the convolution integral

$$E(t) = R(t) * j(t) = \int_{-\infty}^{\infty} R(t_0) j(t - t_0) \, dt_0 \qquad (3.7.3)$$

Irrespective of the domain in which the system is considered, all measurements are made in time and we measure only real quantities. Therefore, $E(t)$ and $j(t)$ are real. It can therefore be concluded from equation (3.7.3) that $R(t)$ is also real. If we require that current j is applied at $t = t_0$, then, due to the causality relation, $E(t) = 0$ for $t < t_0$. From physical necessity it then also follows that $R(t) = 0$ for $t < t_0$.

Taking into consideration the above relations, and substituting in equations (3.7.2) R for f and p for $\bar{f}$, it follows from equation (3.7.2b) that $p(\omega)$ is in general complex (a real p would lead to the trivial case of time-independent resistivity). From equation (3.7.2a), in order for $R(t)$ to be real, the real part of $p(\omega)$ must be an even function and the imaginary part of $p(\omega)$ an odd function, hence

$$p(\omega) = p_{\text{even}}(\omega) + i p_{\text{odd}}(\omega) \qquad (3.7.4)$$

The Fourier transform of an even real function is an even function, and the transformation of an odd imaginary function is an odd function, which yields from equation (3.7.4):

$$R(t) = F^{-1}\{p(\omega)\} = R_{\text{even}}(t) + R_{\text{odd}}(t) \qquad (3.7.5)$$

In order for the relation $R(t) = 0$ for $t < t_0$ be valid, the even and odd parts of R must satisfy the relation

$$R_{\text{odd}}(t) = R_{\text{even}}(t) \, \text{sgn}(t) \qquad (3.7.6)$$

where $\text{sgn}(t)$ is equal to $+1$ for $t > 0$, to 0, for $t = 0$, and to -1 for $t < t_0$. It now follows from the convolution theorem that

$$F\{R_{\text{odd}}(t)\} = F\{R_{\text{even}}(t) * F\{\text{sgn}(t)\}$$

which, by substituting $F\{\text{sgn}(t)\} = 2/i\omega$, yields

$$i p_1(\omega) = p_R(\omega) * \frac{2}{i\omega} \qquad (3.7.7)$$

where p_R is the real (even) part and p_1 the imaginary (odd) part of the resistivity p, which, written in frequency convolution integral form, gives

$$p_1(\omega) = -\frac{1}{\pi} \int_{-\infty}^{\infty} \frac{p_R(\omega_0)}{\omega - \omega_0} \, d\omega_0 \qquad (3.7.8a)$$

where the integral is defined as the Cauchy principal value. Equation (3.7.8) is a Hilbert transform and has an inverse

$$\rho_R(\omega) = \frac{1}{\pi} \int_{-\infty}^{\infty} \frac{\rho_I(\omega_0)}{\omega - \omega_0} d\omega_0 \qquad (3.7.8b)$$

It can be noted from equations (3.7.8) that the real and the imaginary parts of complex resistivity are uniquely related by a Hilbert transform pair.

In a spectral IP survey, the resistivity dispersion is usually analysed on the basis of the magnitude and the phase spectra rather than by using the real and imaginary components. The physical processes that produce the IP are such that the resistivity spectrum is of special character, called the **minimum phase-shift spectrum**, for which the magnitude and the phase angle can also be related by a Hilbert transform pair [28]. Writing ρ in the form

$$\rho(\omega) = \exp[-A(\omega) - i\theta(\omega)] \qquad (3.7.9)$$

the attenuation and phase of $\rho(\omega)$ are related by the following equations similar to equations (3.7.8):

$$\theta(\omega) = -\frac{\omega}{\pi} \int_{-\infty}^{\infty} \frac{A(\omega_0)}{\omega^2 - \omega_0^2} d\omega_0 \qquad (3.7.10a)$$

$$A(\omega) = A(0) + \frac{\omega^2}{\pi} \int_{-\infty}^{\infty} \frac{\theta(\omega_0)}{\omega_0(\omega^2 - \omega_0^2)} d\omega_0 \qquad (3.7.10b)$$

Thus $\theta(\omega)$ can be uniquely determined from $A(\omega)$, whereas both $\theta(\omega)$ and the attenuation $A(0)$ at zero frequency are needed for determining $A(\omega)$.

From the above relations it follows that the entire information contained by a complex resistivity spectrum is in principle represented by either of the spectral components alone. For example, the phase spectrum, if properly sampled, is a sufficient basis for a complete inversion of the whole complex resistivity spectrum and the magnitude spectrum would not in principle add any information to the result. Nevertheless, it is appropriate to analyse both the spectral components, at least for evaluating the quality of the spectral data under the inversion.

It was indicated in Sections 2.4 and 2.5 that in order to represent the charge density in terms of a single field vector (field strength **E** or current density **j**), a sufficiently simple constitutive relation is needed. Such a relation is provided by Ohm's law which, in the frequency domain, expresses the field strength as a product of the current density and resistivity (equation (3.7.1). The resistivity associated with the frequency-domain formulation of IP can thus be readily used in the potential-theoretical formulation of the IP field. In the time domain, the constitutive relation is a convolution integral, as given in equation (3.7.3), which is too complicated to apply as the basis of the potential-theoretical formulation of the IP field. The integral equation solutions for the IP responses are thus in general formulated in the frequency domain. The corresponding time-domain responses are then obtained from the calculated frequency-domain responses by the inverse Fourier transform (equation (3.7.2)).

There seems to be a special case in which the time-domain constitutive relation can be approximated by a simple product, namely when the input current signal has the form of a step function. With this idealization it is possible directly to model the simple time-domain IP response function, known as the **apparent chargeability**.

Assume that the primary current signal has the form of a step function, $I(t<0) = 0$ and $I(t>0) = I_0$. The constitutive relation at a point having resistivity $\rho(\omega)$ is given by equation (3.7.3), where $R(t) = F^{-1}\{\rho(\omega)\}$. Assuming that current I_0 is applied at $t_0 = 0$, it follows that, in equation (3.7.3), $j(t<0) = 0$ and, hence, $R(t<0) = 0$.

If the polarization is weak, we can assume that, since I has the form of a step function, current density j also has this form, hence $j(t<0) = 0$ and $j(t>0) = j_0$. Note that this relation is strictly valid only if $\rho(\omega)$ is homogeneous in the whole model because, in a non-homogeneous model, the current density distribution changes with the changing polarization state. Taking into consideration the above relations, equation (3.7.3) simplifies to

$$E(t) = j_0 \int_0^t R(t_0)\, dt_0 \qquad (3.7.11)$$

Denoting

$$p(t) = \int_0^t R(t_0)\, dt_0 \qquad (3.7.12)$$

the time-domain constitutive relation for a step current excitation appears as an approximation in the form of Ohm's law:

$$E(t) = p(t)\, j_0 \qquad (3.7.13)$$

which enables the direct potential-theoretical treatment of the apparent chargeability.

3.7.3 Modelling of IP anomalies

In Section 3.7.2 we considered the behaviour of resistivity as a function of frequency and time. It was stated that, in order to be able to formulate IP anomalies by the methods of the potential theory, the field strength **E** must be representable as a product of current density **j** and resistivity ρ. This requirement is generally satisfied in the frequency domain where, according to equation (3.7.1)

$$\mathbf{E}(\mathbf{R}; \omega) = \rho(\mathbf{R}; \omega)\, \mathbf{j}(\mathbf{R}; \omega) \qquad (3.7.14)$$

For a step current excitation, **E** and **j** can also, as an approximation, be related in the time domain according to equation (3.7.13) as follows:

$$E(\mathbf{R}; t) = \rho(\mathbf{R}; t) \mathbf{j}(\mathbf{R}) \qquad (3.7.15)$$

In the surface-polarization model, the effect of polarization is explicitly represented by an electric double layer. On the other hand, in the IP polarization model, the polarization is included in the system as a volume property of the medium, the frequency dependence (or, equivalently, time dependence) of resistivity. The electrical sources of the IP anomalies are completely represented by the simple surface charge distributions on the interfaces between media having different resistivity dispersions, and the integral equation modelling of IP can thus be done by using the resistivity modelling equations considered in Section 3.3.

IP modelling is performed by simply substituting the $\rho(\omega)$ or $\rho(t)$ values, properly sampled to meet the requirements of the IP modelling problem, into the integral equations given in Section 3.3. As a consequence of the complex character of resistivity function $\rho(\omega)$ the frequency-domain IP model is complex. These complex resistivity values in turn make the source functions $q_S(\omega)$ or $U(\omega)$ obtained from the integral equations complex and, hence, also the potential and field strength values calculated from these source functions.

IP surveys may also be made by measuring the frequency- (or time-)dependent magnetic field generated by galvanically introduced currents flowing in the ground. This is known as the magnetic IP (MIP) method [29] and is analogous to the magnetometric resistivity (MMR) method presented in Section 3.4. MIP modelling is performed by substituting the resistivity values $\rho(\omega)$ or $\rho(t)$ into the MMR modelling formulas.

3.7.4 Applications

A useful dispersion model as a tool for IP spectral analysis is the Cole–Cole model, originally developed to predict complex dielectric behaviour and subsequently modified to represent the complex resistivity of mineralized rocks [30]. In practice, the *in-situ* resistivity spectra can commonly be described by one or two simple Cole–Cole dispersion models. The formulation of the Cole–Cole model is based on the analogy between alternating-current circuits and mineralized rock models and can be written in the form

$$\rho(\omega) = \rho_0 \left[1 - m \left(1 - \frac{1}{1 + (i\omega \tau)^c} \right) \right] \qquad (3.7.16)$$

where $\rho(\omega)$ is the complex resistivity [Ω m], ρ_0 is the value of ρ at zero frequency, m is the chargeability, c the frequency dependence, τ the time constant [s], ω the angular frequency [rad S^{-1}] and $i^2 = -1$. The characteristic features of a spectrum are thereby solely represented in terms of the four parameters ρ_0, m, c, and τ. It has been observed from *in-situ* measurements [30] that these parameters characterize different structural properties of rocks. In particular, the time constant τ is a crucial parameter in discriminating geological formations having different mineralization textures.

It has been shown by Soininen [31] that the apparent resistivity dispersion due to a body having Cole–Cole dispersive resistivity also follows the form of a Cole–Cole model. It was further shown that the values of the apparent resistivity spectral parameters c_a and τ_a, which are the most important parameters in analysing the Cole–Cole spectra, are very close to the true values, c and τ, of the body resistivity. This means that the Cole–Cole model can be used directly in analysing the apparent resistivity spectra. As a further generalization of the Cole–Cole spectral analysis method, it has been indicated by Soininen [32] that the effect of a polarizable environment on an apparent resistivity spectrum can be approximated by multiplying the spectrum due to the model in an unpolarizable environment by the spectrum of the host medium.

In conventional IP surveys, the polarization response measured is a simple IP quantity, rather than a wide-band resistivity spectrum. The most commonly applied IP quantities in the frequency domain are the frequency effect and the phase shift between the voltage and source current, and, in the time domain, the chargeability. Detailed definitions of these quantities are given by Sumner [26].

Figure 3.15 shows the phase shift anomalies for two polarizable prisms using mid-gradient configuration at frequencies 0.01 Hz, 1 Hz and 10 Hz, when the resistivity spectra in the prisms follow the Cole–Cole model in accordance with equation (3.7.16). Figure 3.16 shows the resistivity phase spectrum of the prism material, and the apparent phase spectrum at point $x = -40$ m, obtained by the measurement arrangement defined in Fig. 3.15. The figures illustrate that to be capable of measuring the IP quantity using optimal frequency values, one should have sufficient prior information concerning the mineralization textures – and thus the time constants – of the geological formations being surveyed. The formation considered in this application has a considerably long time constant τ. Hence, at 0.01 Hz, located close to the maximum of the phase spectrum of Fig. 3.16, the phase anomaly in Fig. 3.15, caused by the mineralization, is strong enough to be separated from the geologically induced noise. In contrast, at 10 Hz, the apparent phase spectrum is attenuated to such an extent that anomalies measured at this frequency cannot be observed with certainty.

3.8 SELF-POTENTIAL

3.8.1 Origin of potentials

Spontaneous ground potentials have been known for over 150 years. For mineral prospecting purposes, they may be classified as **background** potentials and **mineralization** potentials [33].

In geophysical surveying, the most significant background potentials are electrofiltration or streaming potentials and potentials generated by concen-

72 Electrical methods

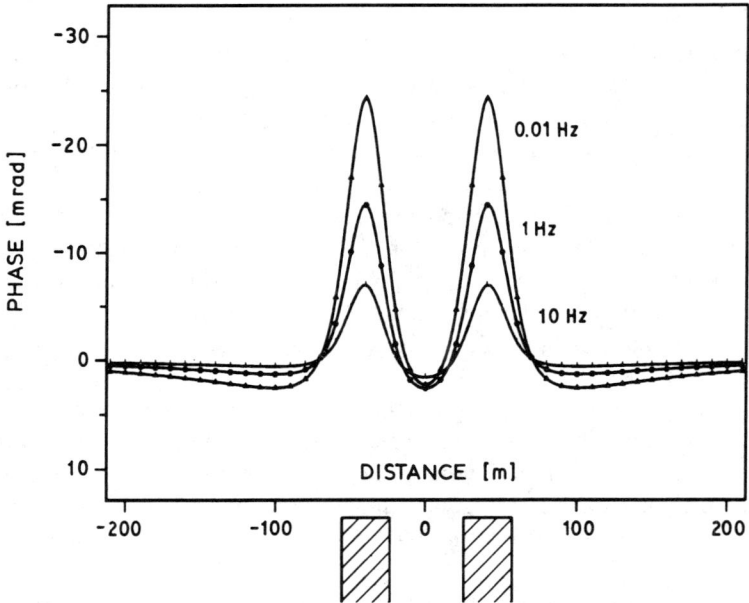

Figure 3.15 Phase angle anomaly for two rectangular prisms at frequencies 0.01 Hz, 1Hz and 10 Hz. Mid-gradient configuration, current electrodes at $x = \pm 1000$ m. Dimensions of prisms 30 m × 150 m × 260 m, depth to top 10 m and distance from each other 50 m. Resistivity of the environment 1000 Ω m; Cole–Cole parameters of the prism's resistivity: $\rho_0 = 200$ Ω m, $m = 0.75$, $\tau = 10$ s and $c = 0.4$.

tration differences in ground water. Their amplitudes vary greatly but are commonly of the order of some tens of millivolts. In geophysical work streaming potential is associated with the flow of water through sand, porous rocks and so on. It seems to be associated with topography in such a way that high ground is more negative than low ground. Streaming potentials have been used to detect leakage spots on submerged slopes of earth-dam water reservoirs, to detect springs and so on, and to study the effect of pumping on the water table [9].

Mineral potentials are of major interest when prospecting with the self-potential (SP) method. They are associated with the metallic sulphides, graphite and sometimes with oxides such as magnetite. The most common mineralization potential anomalies are due to pyrite, chalcopyrite, pyrrhotite, galena and graphite. Amplitudes of mineralization potentials range down to -1 V or even lower.

The microscopic physical and electrochemical processes that produce mineralization potentials are not completely understood, although several theories have been presented to explain them. The most developed electrochemical theory, although not a complete one, is that proposed by Sato and Mooney

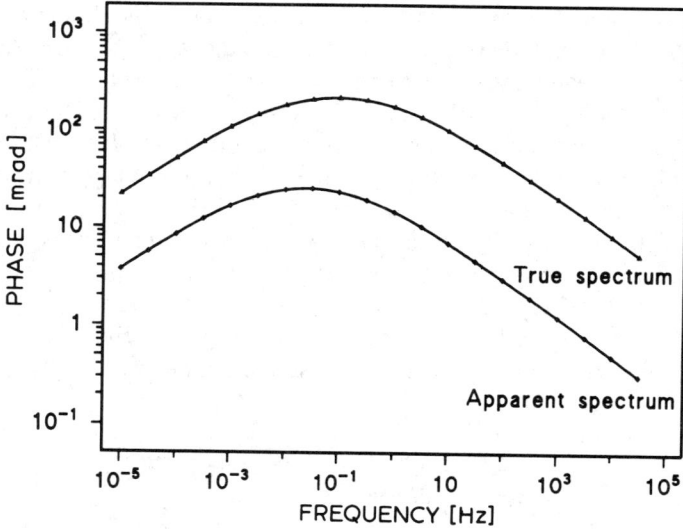

Figure 3.16 Cole–Cole resistivity spectrum of the prism material and apparent resistivity spectrum at point $x = -40$ m of the model defined in Fig. 3.15.

[34]. Their electrochemical theory is based on the postulation of two half-cell reactions, one cathodic above the water table and the other anodic below the water table. In the cathode half-cell, chemical reduction of substances in solution takes place, that is, they receive electrons. In the anode half-cell oxidation of the substances takes place and electrons are lost. An ore body, being a good electronic conductor, serves to transport the electrons from the lower part of the body to the upper part. The intensity of the mineralization potential depends upon the difference in oxidation potential between the solutions at the two half-cells.

In the Sections 3.8.2 and 3.8.3 we shall give integral equation solutions for the mineralization potential and the electrofiltration potential.

3.8.2 Integral equations for mineralization potential

Consider an electrolytically conducting environment with resistivity ρ_2 enveloping an electronically conducting massive sulphide occurrence with resistivity ρ_1, such that $\rho_1 \ll \rho_2$. The spontaneous electrochemical reactions taking place on a sulphide–electrolyte interface generate an electric double layer on the interface [34]. This double layer serves as the primary source of the mineralization potential anomaly (see Fig. 3.17a).

The potential of a double layer distributed on a closed surface is non-zero outside the surface only if the density of the double layer is non-homogeneous. A common situation in mineralization potential models is when above the groundwater table the double layer is negative, that is, the dipole moment is directed into the sulphide body, and below the water table positive. As the ore body is a good conductor, the potential inside the interface is almost constant, but outside, due to the double layer, it is non-homogeneous. The consequent potential gradients drive an ionic current in the electrolyte and an equal amount of electronic return current in the ore body flowing from the lower parts of the model to the upper.

When these conservative currents penetrate the interface, a secondary surface charge layer is generated due to the resistivity contrast between the conductor and the environment (see Fig. 3.17b). In addition, a secondary distribution of the electric double layer is created as the current passes through the surface polarization impedance, as described in Section 3.6. We can con-

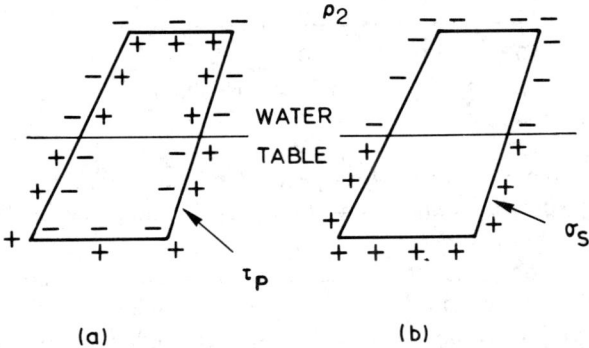

Figure 3.17 Macroscopic sources of mineralization potential. (a) Primary double layer. (b) Secondary surface charge.

clude from the foregoing that the macroscopic physical sources of mineralization potential are the primary and the secondary electric double layers and the secondary simple surface charge distributed on the boundary between the ore body and the environment [35].

The mineralization potential U can now be written according to equations (2.6.3) and (2.5.1) in the form

$$U(\mathbf{R}) = \frac{1}{\varepsilon_0} \int_S \{G'(\mathbf{R}, \mathbf{R}_0) \, \tau(\mathbf{R}_0) + G(\mathbf{R}, \mathbf{R}_0) \, \sigma_S(\mathbf{R}_0)\} \, dS_0 \qquad (3.8.1)$$

where G is Green's function for the potential generated by a simple source and $G' = \partial G/\partial n_0$. τ is the total density of electric double layer on boundary S, which includes both the primary and the secondary double layers, and σ_S is the density of the secondary surface charge.

Taking into consideration that

$$\tau = \tau_P + \tau_S \qquad (3.8.2)$$

where τ_p is the density of primary double layer and τ_S that of the secondary double layer, and applying equations (2.6.6) and (2.5.7), equation (3.8.1) takes the form

$$U(\mathbf{R}) = U_\mathrm{P}(\mathbf{R}) + \int_S \left[\rho_2 G(\mathbf{R}, \mathbf{R}_0) - Z G'(\mathbf{R}, \mathbf{R}_0) \right] j_n(\mathbf{R}_0) \, dS_0 \qquad (3.8.3)$$

where j_n is the normal component of current density on S, Z is the surface polarization impedance, and U_P the primary potential, given by

$$U_\mathrm{P}(\mathbf{R}) = \int_S G'(\mathbf{R}, \mathbf{R}_0) \, \Delta U_\mathrm{P}(\mathbf{R}_0) \, dS_0 \qquad (3.8.4)$$

where $\Delta U_\mathrm{P} = \tau_\mathrm{P}/\varepsilon_0$ is the primary jump in potential across boundary S. We have thus obtained an integral representation for the self-potential model which is analogous to that for the surface-polarization model, except that the primary potentials differ from each other. In the surface-polarization model the primary potential is generated by a known system of current electrodes, whereas in the self-potential model it is generated by a potential jump on boundary S caused by spontaneous electrochemical reactions. Although this potential jump is one of the unknown functions to be interpreted, SP modelling is possible only if it is inserted in (3.8.3) as a known quantity.

To obtain the current density j_n, we develop equation (3.8.3) into an integral equation by moving the calculation point $\mathbf{R}$ to the surface S from the interior side of S. In removing the singular point $\mathbf{R} = \mathbf{R}_0$ from the integrals of equations (3.8.3) and (3.8.4), two additional terms result, one due to the double layer τ_P and the other due to τ_S. Hence, using equation (2.6.4b) and assuming that the interior region of surface S is at constant potential U_C, equations (3.8.3) and (3.8.4) result in the integral equation

$$U_\mathrm{C} = -\frac{1}{2} \Delta U_\mathrm{P}(\mathbf{R}) + \frac{Z}{2} j_n(\mathbf{R}) + \int_S G'(\mathbf{R}, \mathbf{R}_0) \, \Delta U_\mathrm{P}(\mathbf{R}_0) \, dS_0$$

$$+ \int_S [\rho_2 G(\mathbf{R}, \mathbf{R}_0) - Z G'(\mathbf{R}, \mathbf{R}_0)] \, j_n(\mathbf{R}_0) \, dS_0 \qquad (3.8.5)$$

where the surface integrals are defined as principal values.

Both U_C and j_n are unknown and equation (3.8.5) is underdetermined. One supplementary equation is thus needed to solve the unknowns. Here we can again use the continuity equation

$$\int_S j_n(\mathbf{R}_0) \, dS_0 = 0 \qquad (3.8.6)$$

The integral equation pair formed by equations (3.8.5) and (3.8.6) can be solved for j_n by standard numerical methods. Once j_n is known, the potential U can be calculated at any arbitrary point from equation (3.8.3) and (3.8.4).

As the secondary sources of the mineralization SP model are similar to those of the surface-polarization model, the SP model obeys the scaling relations of

76 Electrical methods

the surface-polarization model. Consequently, if SP modelling is done on the laboratory scale, the theoretical description of the model is given by equations (3.8.3)–(3.8.6) in their complete form, as demonstrated by Eskola and Hongisto [35]. On the field scale, however, the contribution of the surface impedance is negligible, and the secondary double layer terms in equations (3.8.3) and (3.8.5) can be disregarded. With this simplification, integral equation (3.8.3) acquires the form:

$$U(\mathbf{R}) = U_P(\mathbf{R}) + \int_S p_2 G(\mathbf{R}, \mathbf{R}_0) j_n(\mathbf{R}_0) \, dS_0 \qquad (3.8.3')$$

Correspondingly, the integral equation (3.8.5) simplifies to

$$U_C = -\frac{1}{2} \Delta U_P(\mathbf{R}) + \int_S G'(\mathbf{R}, \mathbf{R}_0) \Delta U_P(\mathbf{R}_0) \, dS_0$$

$$+ \int_S p_2 G(\mathbf{R}, \mathbf{R}_0) j_n(\mathbf{R}_0) \, dS_0 \qquad (3.8.5')$$

Let us consider by numerical modelling the mutual contributions of the primary double layer and secondary surface charge distributions to the total SP anomaly. The calculations simulate the field-scale model, that is to say, the effect of the surface impedance has not been taken into account. Figure 3.18 represents the primary, secondary and total SP anomalies of a prism associated with three different water table levels. Below the water table, the prism surface was given a constant primary potential jump of $+1$ V, and above the water table, zero. It can be seen that the total anomaly becomes stronger and sharper as the groundwater table rises, which indicates that the rising water table brings the effective anomaly sources closer to the measurement level. Both absolutely and relatively, the effect of the secondary charge is strongest in the case of the medium groundwater height, which is depicted in Fig. 3.18b. This is probably due to the fact that, in the case of highest groundwater level, the positive image charge distributed on the ground surface hinders upward current flow, which weakens the current density and the associated negative charge density on the upper face of the prism.

It was noted previously that the primary source of SP is known to be located on the surface of the ore in the form of a double layer whose density is one of the unknown quantities to be interpreted. Hence, the interpretation of SP models is in practice more difficult than that for other electrical models, for which the location and strength of the primary source are known constraints to the problem. It is thus important to constrain the possible distribution modes of primary sources on geological or theoretical grounds. One theoretical constraining principle is the equivalence of sources, which means different source distributions that generate similar primary potentials. For example, the potential of a homogeneous double layer distributed on a closed surface is zero outside the surface. From the superposition principle it then follows that the

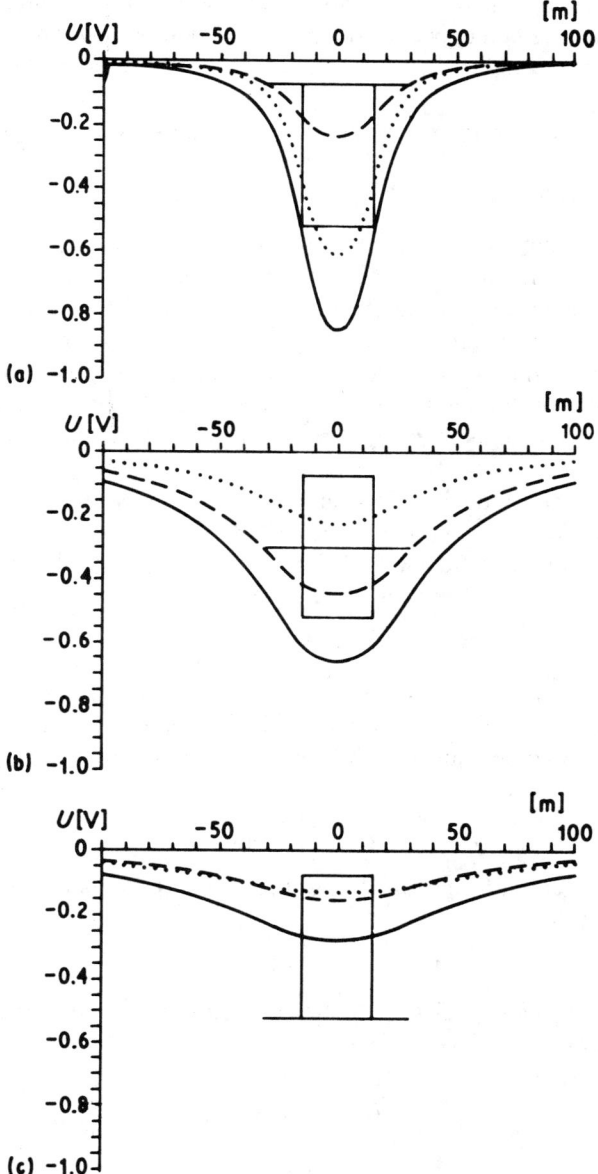

Figure 3.18 Primary (dotted line), secondary (dashed line), and total (solid line) mineralization potential anomaly due to $2\frac{1}{2}$-dimensional prism with strike length 500 m. Groundwater level at (a) 10 m; (b) 40 m; (c) 70 m.

78   *Electrical methods*

SP anomaly of a body is always the same when the primary source is composed of two parts, separated by a horizontal surface (for example, the water table), both having homogeneous double layers with an invariable difference in layer density.

3.8.3 Integral equations for electrofiltration potential

The boundary-value problem

Electrofiltration potentials generally arise in moderately conductive geological formations, and therefore the integral equation solution based on the equipotential condition and applied in connection with the mineralization potentials cannot be applied. In the following we shall define the boundary-value problem for the electrofiltration potential model and give two alternative integral equation solutions for it. Since these solutions have not so far been programmed for computers, they are presented without numerical applications.

Consider an earth model as illustrated in Fig. 2.1, in which the boundary S is assumed to be a discontinuity surface of both the resistivity ρ and the streaming potential coefficient C. The total electric potential φ for the electrofiltration model is defined as follows [36]:

$$\varphi = U + CP \qquad (3.8.7)$$

where P is pressure, U is the electric potential associated with conduction currents, and CP is the streaming potential. The total current density $\mathbf{j}$, comprising both the conduction and the convection currents, is thus represented by Ohm's law

$$\mathbf{j} = -\frac{1}{\rho}\nabla\varphi$$

The current is stationary and laminar, so that $\Delta \cdot \mathbf{j} = 0$, and potential φ satisfies Laplace's equation

$$\nabla^2 \varphi = 0 \qquad (3.8.8)$$

in regions V_1, V_2 and V_3. The potential φ is taken to satisfy the following boundary conditions:

$$\frac{1}{\rho_1}\frac{\partial \varphi_1}{\partial n} = \frac{1}{\rho_2}\frac{\partial \varphi_2}{\partial n}$$

$$U_1 = U_2 \qquad (3.8.9)$$

$$P_1 = P_2$$

on the surface S. $\mathbf{n}$ is the outward unit normal vector of surfaces S and A. From the latter two conditions and equation (3.8.7) it follows that

$$\varphi_2 - \varphi_1 = (C_2 - C_1)P \qquad (3.8.10)$$

on the surface S.

On the outer boundary A of region V_2, φ satisfies the boundary conditions:

$$\frac{1}{\rho_2}\frac{\partial \varphi_2}{\partial n} = \frac{1}{\rho_3}\frac{\partial \varphi_3}{\partial n}$$

$$\varphi_2 = \varphi_3 \qquad (3.8.11)$$

Laplace's equation (3.8.8) and the boundary conditions (3.8.9)–(3.8.11) define the boundary-value problem for the total electric potential φ, for which we shall derive below two alternative integral equation solutions. The first of these is a physical formulation for the charge density, and the second is a non-physical formulation for potential as being due to fictitious double sources.

Integral equation for charge density

It can be concluded from the boundary condition (3.8.10) that the physical primary source for the electrofiltration potential is the electrical double layer τ_p, which is obtained from equation (3.8.10) as follows:

$$\tau_P = \varepsilon_0 (\varphi_2 - \varphi_1) = \varepsilon_0 (C_2 - C_1) P \qquad (3.8.12)$$

A necessary and sufficient condition to ensure the primary convection current flow, which is the cause of the electric double layer τ_p, is that P varies along the boundary S between regions having different streaming potential coefficients C.

The potential jump associated with the primary double layer τ_P generates a primary electric field which makes conduction currents flow in the model. When these currents flow across the boundary S, a surface charge distribution σ_S is generated on S due to the resistivity discontinuity. The total potential of the model is thus the sum of the potentials due to the primary double layer τ_p and the secondary surface charge σ_S.

The potential in the model under consideration can be written by analogy with equation (3.3.8) as follows:

$$\varphi(\mathbf{R}) = \varphi_P(\mathbf{R}) + \int_S G(\mathbf{R},\mathbf{R}_0)\, q_S(\mathbf{R}_0)\, dS_0 \qquad (3.8.13)$$

where the primary potential φ_P is obtained, using equation (3.8.12) with $\Delta\varphi = \varphi_2 - \varphi_1$, from the following equation:

$$\begin{aligned}\varphi_P(\mathbf{R}) &= \frac{1}{\varepsilon_0}\int_S G'(\mathbf{R},\mathbf{R}_0)\, \tau_P(\mathbf{R}_0)\, dS_0 \\ &= \int_S G'(\mathbf{R},\mathbf{R}_0)\, \Delta\varphi(\mathbf{R}_0)\, dS_0 \\ &= (C_2 - C_1)\int_S G'(\mathbf{R},\mathbf{R}_0)\, P(\mathbf{R}_0)\, dS_0 \end{aligned} \qquad (3.8.14)$$

80   Electrical methods

Green's function for the double source is $G' = \partial G/\partial n_0$, where G is Green's function for the potential generated by a simple source, defined by Poisson's equation (3.3.17) and the boundary conditions (3.3.18). Further, G must be regular at infinity, that is, it must approach zero as $1/R$, as R approaches infinity.

In equation (3.8.13), the charge density q_S is unknown, whereas τ_P, and thus $\Delta \varphi$ and P, are assumed to be known primary source distributions.

To obtain q_S, equation (3.8.13) is developed into an integral equation by moving the calculation point $\mathbf{R}$ to the surface S and by differentiating both sides in the direction of normal $\mathbf{n}$. Removing the contribution of the singular point $\mathbf{R} = \mathbf{R}_0$ in accordance with equation (2.5.4), we obtain the following integral equation:

$$\frac{\rho_2 + \rho_1}{2(\rho_2 - \rho_1)} q_S(\mathbf{R}) = \frac{\partial \varphi_P(\mathbf{R})}{\partial n} - \int_S \frac{\partial G(\mathbf{R}, \mathbf{R}_0)}{\partial n} q_S(\mathbf{R}_0) \, \mathrm{d}S_0 \qquad (3.8.15)$$

After equation (3.8.15) has been solved for q_S, the SP anomaly φ can be calculated at any arbitrary point of the regions V_1 and V_2.

Note that after differentiation, Green's function in the primary term of integral equation (3.8.15) is $G'' = \partial^2 G/\partial n_0 \partial n$, which is a rapidly varying function of order $1/R^3$. Fortunately, this function appears only in the primary term and not in the integral, which permits the numerical solution of equation (3.8.15), if care is taken that the primary field is integrated with sufficient accuracy.

Integral equation for potential

We shall now derive for the electrofiltration potential problem an integral equation solution in which the potential is generated by fictitious double sources. The formulation is analogous to that given in Section 3.3.3 for the resistivity problem.

Consider the boundary value problem defined for the earth model of Fig. 2.1 by Laplace's equation (3.8.8) and the boundary conditions (3.8.9)–(3.8.11). To convert this boundary-value problem into an integral equation, we define Green's function G by Poisson's equation (3.3.17) and boundary conditions consisting of equation (3.3.18) and the regularity condition at infinity.

Let us substitute functions φ and G into Green's second identity and apply the identity first in the region V_2 such that both $\mathbf{R}$ and $\mathbf{R}_0$ are in V_2.

Taking into consideration that the surface integral over A vanishes due to the boundary conditions (3.8.11) and (3.3.18), the identity acquires the following form:

$$\int_{V_2} (G \nabla^2 \varphi_2 - \varphi_2 \nabla^2 G) \, \mathrm{d}V = \int_S \left(-G \frac{\partial \varphi_2}{\partial n} + \varphi_2 \frac{\partial G}{\partial n} \right) \mathrm{d}S \qquad (3.8.16)$$

Self-potential

Substituting the partial differential equations (3.8.8) and (3.3.17), equation (3.8.16) simplifies to

$$\varphi_2 = \int_S \left(-G \frac{\partial \varphi_2}{\partial n} + \varphi_2 \frac{\partial G}{\partial n} \right) dS \quad (3.8.17)$$

We next apply the identity to region V_1, keeping point $\mathbf{R}_0$ in region V_2. Again substituting equation (3.8.8) and (3.3.17), the identity yields

$$0 = \int_S \left(G \frac{\partial \varphi_1}{\partial n} - \varphi_1 \frac{\partial G}{\partial n} \right) dS \quad (3.8.18)$$

Multiplying equation (3.8.18) by ρ_2/ρ_1 and adding the equation thus obtained to equation (3.8.17), and substituting the first of boundary conditions (3.8.9), we obtain

$$\varphi_2 = \int_S \frac{\partial G}{\partial n} \left(\varphi_2 - \frac{\rho_2}{\rho_1} \varphi_1 \right) dS \quad (3.8.19)$$

Substituting from equation (3.8.14) $\varphi_1 = \varphi_2 - \Delta\varphi$, where $\Delta\varphi$ is the known primary potential jump, replacing $\mathbf{R}_0$ with $\mathbf{R}$, and using the reciprocity relation $G(\mathbf{R}_0, \mathbf{R}) = G(\mathbf{R}, \mathbf{R}_0)$, we obtain for the potential φ_2

$$\varphi_2(\mathbf{R}) = \varphi_{P2}(\mathbf{R}) + \frac{\rho_1 - \rho_2}{\rho_1} \int_S G'(\mathbf{R}, \mathbf{R}_0) \varphi_2(\mathbf{R}_0) \, dS_0 \quad (3.8.20)$$

in region V_2. Green's function for the double source is $G' = \partial G/\partial n_0$, and the (fictitious) primary potential φ_{P2} is given by

$$\varphi_{P2}(\mathbf{R}) = \frac{\rho_2}{\rho_1} \int_S G'(\mathbf{R}, \mathbf{R}_0) \Delta\varphi(\mathbf{R}_0) \, dS_0 \quad (3.8.21)$$

To obtain the corresponding representation for the potential in region V_1, we keep point $\mathbf{R}_0$ in V_1 and apply Green's second identity to regions V_1 and V_2. Proceeding as above, we obtain for potential φ_1

$$\varphi_1(\mathbf{R}) = \varphi_{P1}(\mathbf{R}) + \frac{\rho_1 - \rho_1}{\rho_2} \int_S G'(\mathbf{R}, \mathbf{R}_0) \varphi_1(\mathbf{R}_0) \, dS_0 \quad (3.8.22)$$

in region V_1. The primary potential is now given as

$$\varphi_{P1}(\mathbf{R}) = \frac{\rho_1}{\rho_2} \int_S G'(\mathbf{R}, \mathbf{R}_0) \Delta\varphi(\mathbf{R}_0) \, dS_0$$

To find the unknown potential functions φ_1 and φ_2 on the surface S, we first develop equation (3.8.20) into an integral equation by moving the calculation point $\mathbf{R}$ in region V_2 to the surface S. The integrals representing the primary and secondary potentials each generate an additional term due to the singular point $\mathbf{R} = \mathbf{R}_0$, both of which are taken into consideration in accordance with equation (2.6.4a) so that

82   *Electrical methods*

$$\frac{\rho_1+\rho_2}{2\rho_1}\varphi_2(\mathbf{R}) = \frac{\rho_2}{2\rho_1}\Delta\varphi(\mathbf{R}) + \frac{\rho_2}{\rho_1}\int_S G'(\mathbf{R}, \mathbf{R}_0)\Delta\varphi(\mathbf{R}_0)\,dS_0$$

$$+ \frac{\rho_1-\rho_2}{\rho_1}\int_S G'(\mathbf{R}, \mathbf{R}_0)\varphi_2(\mathbf{R}_0)\,dS_0 \qquad (3.8.24)$$

where the integrals are defined as the principal values.

The corresponding integral equation for calculating the potential φ_1 on the surface S can be obtained either from equation (3.8.22) by following an approach similar to that above, or simply by substituting in equation (3.8.24) the expression $\varphi_2 = \varphi_1 - \Delta\varphi$. Both procedures result in the equation

$$\frac{\rho_1+\rho_2}{2\rho_2}\varphi_1(\mathbf{R}) = -\frac{\rho_1}{2\rho_2}\Delta\varphi(\mathbf{R}) + \frac{\rho_1}{\rho_2}\int_S G'(\mathbf{R}, \mathbf{R}_0)\Delta\varphi(\mathbf{R}_0)\,dS_0$$

$$+ \frac{\rho_1-\rho_2}{\rho_2}\int_S G'(\mathbf{R}, \mathbf{R}_0)\varphi_1(\mathbf{R}_0)\,dS_0 \qquad (3.8.25)$$

Note that to obtain the potential φ for the whole model $V_1 + V_2$, only one of the integral equations (3.8.24) and (3.8.25) need be solved; as soon as one of the functions φ_1 and φ_2 on S has been obtained, the other can be calculated by simply applying the relation $\varphi_2 - \varphi_1 = \Delta\varphi$.

When φ_1 and φ_2 on surface S are known, the potential can be calculated at any arbitrary point in regions V_1 and V_2 from equations (3.8.22) and (3.8.20), respectively.

3.9 ELECTRICAL ANISOTROPY

So far we have only considered electrically isotropic models, in which the relation between the electric field and current density is expressed by a simple scalar coefficient. However, many rocks such as shales, slates and laminated ores possess considerable electrical anisotropy. In this section we shall consider the electrostatic properties associated with electrical anisotropy, and give integral equation solutions for the potential in two anisotropic resistivity models, one of which is composed of a perfect conductor and the other of an anisotropic body, located in an anisotropic space.

3.9.1 Definition

When the properties of the medium do not depend on the direction in the medium along which they are measured, the medium is said to be **isotropic**. Thus, anisotropy is defined as the variation of a particular property with direction.

In the following, we shall consider anisotropic resistivity models, in which it is assumed that the constitutive relation between the electric field $\mathbf{E}$ and current density **j** is linear and depends upon the direction of current flow in the medium. It is most convenient to express this anisotropic Ohm's law by means of a tensor, which can be written, using Cartesian coordinates, in the form:

$$\mathbf{E} = \tilde{\rho}\,\mathbf{j} \tag{3.9.1}$$

where $\tilde{\rho}$ is the resistivity tensor. Written in component form, equation (3.9.1) is given as:

$$\mathbf{E}_x = \rho_{xx}\mathbf{j}_x + \rho_{xy}\mathbf{j}_y + \rho_{xz}\mathbf{j}_z$$
$$\mathbf{E}_y = \rho_{yx}\mathbf{j}_x + \rho_{yy}\mathbf{j}_y + \rho_{yz}\mathbf{j}_z$$
$$\mathbf{E}_z = \rho_{zx}\mathbf{j}_x + \rho_{zy}\mathbf{j}_y + \rho_{zz}\mathbf{j}_z \tag{3.9.2}$$

The inverse of equation (3.9.1) is

$$\mathbf{j} = \tilde{g}\,\mathbf{E} \tag{3.9.3}$$

where the conductivity tensor is the inverse of the resistivity tensor, $\tilde{g} = \tilde{\rho}^{-1}$. Equation (3.9.3) can be expressed in component form as follows:

$$\mathbf{j}_x = g_{xx}\mathbf{E}_x + g_{xy}\mathbf{E}_y + g_{xz}\mathbf{E}_z$$
$$\mathbf{j}_y = g_{yx}\mathbf{E}_x + g_{yy}\mathbf{E}_y + g_{yz}\mathbf{E}_z$$
$$\mathbf{j}_z = g_{zx}\mathbf{E}_x + g_{zy}\mathbf{E}_y + g_{zz}\mathbf{E}_z \tag{3.9.4}$$

It is apparent from equations (3.9.2) and (3.9.4) that the relation of **E** and **j** in an anisotropic medium is characterized by nine coefficients represented collectively by ρ_{ij}. However, a fundamental property of the current conduction system is that the resistivity and the conductivity tensors are symmetric, that is, $\rho_{ij} = \rho_{ji}$ and $g_{ij} = g_{ji}$. Due to this symmetry, the three reference axes can be orientated such that only three independent coefficients are required to describe the relation between **E** and **j**. Denoting these values, known as the **principal values**, by ρ_a, ρ_b and ρ_c, for resistivity, and by g_a, g_b and g_c for conductivity, the relation between **E** and **j** can be expressed in the system having principal axes a, b, c as follows:

$$\mathbf{E} = \rho_a\mathbf{j}_a + \rho_b\mathbf{j}_b + \rho_c\mathbf{j}_c \tag{3.9.5}$$

and

$$\mathbf{j} = g_a\mathbf{E}_a + g_b\mathbf{E}_b + g_c\mathbf{E}_c \tag{3.9.6}$$

where $g_a = 1/\rho_a, g_b = 1/\rho_b$ and $g_c = 1/\rho_c$.

3.9.2 Physical significance

The theoretical structure of an anisotropic model is considerably more complicated than that of an isotropic model [37]. For example, in an anisotropic whole-space the equipotential surfaces of a point current source are stretched into ellipsoids, whereas in an isotropic space they are spheres. In a space containing resistivity boundaries, the complexity of the anisotropic model further increases relative to that of an isotropic model.

The contribution of anisotropy is particularly significant in the application of integral equations, because whereas Green's function is independent of resistivity for an isotropic model, this is not the case for an anisotropic model. This makes the Green's functions associated with physically formulated integral equations both formally and conceptually very problematic.

Let us examine the electrostatic properties of an anisotropic system by means of a simple model, in which current I is emitted into an anisotropic whole-space through a point electrode. Current density $\mathbf{j}$ obeys the conservation relation

$$\nabla \cdot \mathbf{j} = I\delta(\mathbf{R} - \mathbf{R}_P) \tag{3.9.7}$$

where δ is Dirac's delta function and $\mathbf{R}_P$ is the earthing point of the current electrode. Applying equation (3.9.3) and expressing the electric field as a gradient of the scalar potential U, $\mathbf{E} = -\nabla U$, we obtain for the potential U:

$$\nabla \cdot (\widetilde{g}\,\nabla U) = -I\delta(\mathbf{R} - \mathbf{R}_P) \tag{3.9.8}$$

For a whole-space, the only boundary condition satisfied by U is the regularity condition at infinity: U approaches zero as $1/R$, as R approaches infinity. Orientating the system of coordinates so that the coordinate axes x, y, z coincide with the principal axes of anisotropy, placing the point $\mathbf{R}_P$ at the origin and remembering that $\widetilde{g}^{-1} = \widetilde{\rho}$, the solution for the boundary-value problem under consideration can be written in the form

$$U = \frac{\rho I}{4\pi \mathscr{R}} \tag{3.9.9}$$

where

$$\mathscr{R} = [(\widetilde{\rho}/\rho)\mathbf{R} \cdot \mathbf{R}]^{\frac{1}{2}} = (\alpha x^2 + \beta y^2 + \gamma z^2)^{\frac{1}{2}} \tag{3.9.10}$$

$$\alpha = \rho_x/\rho$$

$$\beta = \rho_y/\rho$$

$$\gamma = \rho_z/\rho \tag{3.9.11}$$

$$\rho = (\rho_x \rho_y \rho_z)^{\frac{1}{3}}$$

$$g = (g_x g_y g_z)^{\frac{1}{3}} \tag{3.9.12}$$

g_x, g_y, and g_z refer to conductivity and $\rho_x = 1/g_x$, $\rho_y = 1/g_y$ and $\rho_z = 1/g_z$ refer to resistivity along the principal axes of anisotropy x, y and z.

We see that the expression for potential given in equation (3.9.9), like that for the potential in an isotropic space, comprises a point charge magnitude $q = \varepsilon_0 I \rho$, and a geometric factor $1/4\pi \mathscr{R}$. However, the geometric factor for the anisotropic space depends not only on the space coordinates but also on the conductivity (or resistivity) in such a way that the coordinates have as coefficients dimensionless functions of the principal values of conductivity. These coefficients are responsible for transforming the equipotential surfaces into ellipsoids whose axes coincide with the principal axes of anisotropy.

The field strength $\mathbf{E}$ can be obtained from equation (3.9.9) by the relation $\mathbf{E} = -\nabla U$, hence

$$\mathbf{E} = \frac{I\tilde{\rho}\mathbf{R}}{4\pi \mathscr{R}^3} = \frac{\rho I}{4\pi \mathscr{R}^3}(\alpha \mathbf{x} + \beta \mathbf{y} + \gamma \mathbf{z}) \qquad (3.9.13)$$

Equation (3.9.13) shows that the lines of force of $\mathbf{E}$ are straight along the principal axes of anisotropy but in other directions bend from high towards low conductivity. Substituting Ohm's law (equation (3.9.5)) into equation (3.9.13), we see that the current density lines are straight and spread radially from the current electrode.

We shall now analyse the physical causes for the deformation of an electric field. Since the only sources of a static electric field are the charges, the field bends only if there are volume charge distributions in the environment of the point current source. The charge density σ can be obtained from Gauss's law, $\sigma = \varepsilon_0 \nabla \cdot \mathbf{E}$, as follows:

$$\sigma = \frac{\varepsilon_0 \rho I}{4\pi \mathscr{R}^5}[\beta + \gamma - 2\alpha)\alpha x^2 + (\alpha + \gamma - 2\beta)\beta y^2 + (\alpha + \beta - 2\gamma)\gamma z^2] \qquad (3.9.14)$$

The point current source in an anisotropic medium is thus enveloped by a volume charge distribution whose density decreases outwards from the source point as $1/\mathscr{R}^3$.

Along a principal axis, for example the x-axis:

$$\sigma_{y=z=0} \sim \frac{\beta + \gamma - 2\alpha}{(\alpha x^2)^{3/2}} \qquad (3.9.15)$$

from which it follows that $\sigma \_ 0$ according as $\rho_y + \rho_z \_ 2\rho_x$. That is, the charge is positive along the shortest principal axis of resistivity and negative along the longest axis.

For a more detailed analysis of the charge distribution, the surfaces of zero charge density are determined. Inserting $\sigma = 0$ in equation (3.9.14), we obtain

$$P(x, y, z) = Ax^2 + By^2 + Cz^2 = 0 \qquad (3.9.16)$$

where $A = (\beta + \gamma - 2\alpha)\alpha$, $B = (\alpha + \gamma - 2\beta)\beta$, and $C = (\alpha + \beta - 2\gamma)\gamma$. The coordinate axes x, y, z are chosen such that $\rho_x \geqslant \rho_y \geqslant \rho_z$. Consequently, $A < 0$ and $C > 0$ hold always, whereas the sign of B varies depending on the relative length of the principal axis ρ_y.

When $B < 0$, the surface $P = 0$ represents an elliptic double cone with the vertex at the origin and the z-axis as its axis. The charge density is positive within the cones and negative outside them. When $B > 0$, the surface $P = 0$ is an elliptical double cone with the vertex at the origin and the x-axis as its axis. Within the cones the charge is negative but outside them positive. When B approaches zero these double cones stretch along the y-axis in such a way that, at $B = 0$, the surface $P = 0$ represents two planes that intersect along the y-axis. The charge is negative in the regions of space containing the x-axis, and positive in their complementary regions.

We shall finally calculate the net volume charge Q_V contained in the space surrounding the point charge at the origin. Applying Gauss's law (equation (2.1.7)) and the divergence theorem, we obtain

$$Q_V = \varepsilon_0 \int_{V-v} \nabla \cdot \mathbf{E} \, dV = \varepsilon_0 \int_{S-s} E_n \, dS \qquad (3.9.17)$$

Volume V is extended over the whole space, and v is a small volume containing the point charge at the origin. Surface S encloses the entire volume V, and s encloses the small volume v. Transforming into the system of spherical polar coordinates by

$$x = R \sin\theta \cos\varphi$$

$$y = R \sin\theta \sin\varphi$$

$$z = R \cos\theta$$

$$dS = R^2 \sin\theta \, d\theta \, d\varphi$$

$$E_n = E_R = -\partial U/\partial R$$

we obtain for the volume charge Q_V:

$$Q_V = \frac{\varepsilon_0 \rho I}{4\pi} (\lim_{R \to \infty} - \lim_{r \to 0}) \int_0^{2\pi} d\varphi \int_0^{\pi} F(\theta, \varphi) \sin\theta \, d\theta \qquad (3.9.18)$$

$$F = [(\alpha \cos^2\varphi + \beta \sin^2\varphi) \sin^2\theta + \gamma \cos^2\theta]^{-\frac{1}{2}}$$

where R is the radius of surface S and r that of surface s.

Since the integrand in equation (3.9.18) is independent of variable R, its limits as R approaches infinity and r approaches zero are equal, and hence, $Q_V = 0$.

In summary, for an anisotropic medium, a volume charge distributed outside the current earthing point bends the lines of force of the electric field and stretches the equipotential surfaces into ellipsoids. The charge distribution is controlled by the relative lengths of the principal axes of resistivity. The charge density varies as a function of direction, being positive in the environment of short principal axes of resistivity and negative along the long axes, and its absolute value decreases outwards from the current earthing point as $1/R^3$. The variation of the volume charge in sign and intensity is such that the net volume charge is zero, and the total charge in the space is thus equal to the charge concentrated at the current earthing point.

In an isotropic model, secondary charges occur only in regions of resistivity variation. In contrast, in an anisotropic model, volume charges appear which are characteristic of the resistivity anisotropy and extend over the whole model, including its homogeneous regions. In particular, a surface charge generated on a boundary between two different anisotropy structures is associated, on either side of the boundary, with volume charges that differ from each other. In applying an integral equation formulated on a physical basis, Green's function should contain information on the position of the boundary S and on the resistivity tensors in the regions on either side of it. The difficulties in determining Green's function may then be comparable to those in the analytical solution of the problem.

In some cases, Green's function can be determined by formulating the integral equation solutions in terms of fictitious source functions. In these formulations, both the source and the Green's functions have different forms in regions having different resistivity tensors. As a result, Green's function must only contain the anisotropy information of the region in which it operates, and the effect of the volume charges associated with the anisotropy in other regions of the model is included in the source function.

3.9.3 Integral equation for perfect conductor in an anisotropic environment

Let us consider, after [38], a model in which a perfectly conducting body is located in an environment having anisotropic conductivity. The possible electrical anisotropy in the body does not have any significant effect on the electric

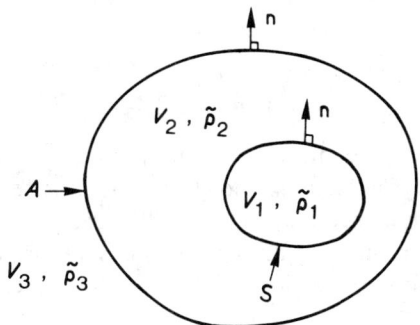

Figure 3.19 Anisotropic earth model.

88   *Electrical methods*

potential in the model if all the three principal conductivities in the body are sufficiently high in relation to those in the environment, thus ensuring that the body is at constant potential.

The model is illustrated in Fig. 3.19, where $\tilde{\rho}_1 = 0$. S is the surface bounding the conductive body V_1 and its environment V_2 and the surface A symbolizes the boundaries of the space surrounding V_1. A primary current I is emitted into the model through a point electrode located at point $\mathbf{R}_P$ in region V_2.

The partial differential equation forming the basis of the boundary-value problem under consideration is the conservation relation of current:

$$\nabla \cdot \mathbf{j} = I\delta(\mathbf{R} - \mathbf{R}_P) \tag{3.9.19}$$

where $\mathbf{j}$ is the current density and δ Dirac's delta function. Using Ohm's law, $\mathbf{j} = \tilde{g}\,\mathbf{E}$, and expressing the field strength in terms of a scalar potential, $\mathbf{E} = -\nabla U$, equation (3.9.19) results in

$$\nabla \cdot (\tilde{g}\,\nabla U) = -I\delta(\mathbf{R} - \mathbf{R}_P) \tag{3.9.20}$$

The solution of equation (3.9.20) must satisfy the boundary conditions

$$U_2 = U_3$$

$$(\tilde{g}_2 \nabla U_2) \cdot \mathbf{n} = (\tilde{g}_3 \nabla U_3) \cdot \mathbf{n} \tag{3.9.21}$$

on A. Further, U must be regular at infinity, that is, approach zero as $1/R$, as R approaches infinity.

To convert the boundary-value problem defined by equations (3.9.20) and (3.9.21) into an integral equation, use is made of Green's function, defined by Poisson's equation

$$\nabla \cdot (\tilde{g}\,\nabla \mathscr{G}) = -g\delta(\mathbf{R} - \mathbf{R}_0) \tag{3.9.22}$$

where $g = (g_a g_b g_c)^{\frac{1}{3}}$, and g_a, g_b, and g_c are the principal values of conductivity $\tilde{g}$. The solution of equation (3.9.22) must satisfy the boundary conditions

$$\mathscr{G}_2 = \mathscr{G}_3 \tag{3.9.23}$$

$$(\tilde{g}_2 \nabla \mathscr{G}_2) \cdot \mathbf{n} = (\tilde{g}_3 \nabla \mathscr{G}_3) \cdot \mathbf{n}$$

on A, and the regularity condition at infinity.

We next substitute functions U and $\mathscr{G}$ into Green's second identity generalized for an anisotropic medium, and apply the identity to region V_2 assuming that both $\mathbf{R}$ and $\mathbf{R}_0$ are in V_2. Taking into consideration the symmetry of the conductivity tensor $\tilde{g}$ (see Section 3.9.1), the identity can be written in the following form [39]:

$$\int_{V_2} \left[\mathscr{G}_2 \nabla \cdot (\tilde{g}_2 \nabla U_2) - U_2 \nabla \cdot (\tilde{g}_2 \nabla \mathscr{G}_2) \right] dV$$

$$= \int_S \left[-\mathscr{G}_2(\tilde{g}_2 \nabla U_2) \cdot \mathbf{n} + U_2(\tilde{g}_2 \nabla \mathscr{G}_2) \cdot \mathbf{n} \right] dS$$

$$+ \int_A \left[\mathscr{G}_2(\tilde{g}_2 \nabla U_2) \cdot \mathbf{n} - U_2(\tilde{g}_2 \nabla \mathscr{G}_2) \cdot \mathbf{n} \right] dA \qquad (3.9.24)$$

Applying the identity to region V_3 while keeping the point $\mathbf{R}_0$ in V_2, adding the equation thus obtained to equation (3.9.24), substituting equations (3.9.20)–(3.9.23) and simplifying the notation by dropping the subscripts, equation (3.9.24) reduces to

$$-\mathscr{G} I + g U = \int_S \left[-\mathscr{G}(\tilde{g} \nabla U) \cdot \mathbf{n} + U(\tilde{g} \nabla \mathscr{G}) \cdot \mathbf{n} \right] dS \qquad (3.9.25)$$

We now assume that, as a consequence of high conductivity within it, region V_1 is at a constant potential $U = U_C$. Since the potential is assumed to be continuous in passing through surface S, the second term in the surface integral of equation (3.9.25) becomes

$$U_C \int_S (\tilde{g} \nabla \mathscr{G}) \cdot \mathbf{n} \, dS = 0 \qquad (3.9.26)$$

This can be proved by using the divergence theorem to transform the surface integral in equation (3.9.26) to a volume integral over region V_1, and by substituting equation (3.9.22) into the integral thus obtained. Another way of showing the validity of equation (3.9.26) is to consider that the integrand represents the normal component of current density on surface S generated by a point current source located at point $\mathbf{R}_0$, and the integral thus simply represents the conservation relation for current.

Substituting Ohm's law, $\mathbf{j} = \tilde{g} \mathbf{E}$, interchanging the points $\mathbf{R}$ and $\mathbf{R}_0$, and applying the reciprocity relation $\mathscr{G}(\mathbf{R}_0, \mathbf{R}) = \mathscr{G}(\mathbf{R}, \mathbf{R}_0)$, equation (3.9.25) gives potential U in the following form:

$$U(\mathbf{R}) = U_P(\mathbf{R}) + \rho \int_S \mathscr{G}(\mathbf{R}, \mathbf{R}_0) j_n(\mathbf{R}_0) \, dS_0 \qquad (3.9.27)$$

$$U_P(\mathbf{R}) = \rho \mathscr{G}(\mathbf{R}, \mathbf{R}_P) I(\mathbf{R}_P) \qquad (3.9.28)$$

where $\rho = 1/g = (\rho_a \rho_b \rho_c)^{\frac{1}{3}}$, and ρ_a, ρ_b and ρ_c are the principal values of resistivity.

To solve the unknown normal component of current density j_n, an integral equation is formed from equation (3.9.27) by moving the calculation point $\mathbf{R}$ to the surface S. The Green's function of equation (3.9.27) is of the form $1/\mathscr{R}$, thus the integral converges and the integral equation obtained is of the same form as equation (3.9.27). Assuming that, due to the high conductivity in region V_1, the potential is constant, $U = U_C$ on the surface S, the integral equation can be written as:

90   *Electrical methods*

$$U_C = U_P(\mathbf{R}) + \rho \int_S \mathscr{G}(\mathbf{R}, \mathbf{R}_0) j_n(\mathbf{R}_0) \, dS_0 \qquad (3.9.29)$$

Since both U_C and j_n are unknown, equation (3.9.27) is underdetermined, and hence we use as a supplementary equation the conservation relation for current:

$$\int_{S(I)} j_n \, dS = I \qquad (3.9.30a)$$

for surface $S(I)$ enclosing a source emitting current I, and

$$\int_{S(\text{no } I)} j_n \, dS = 0 \qquad (3.9.30b)$$

for surfaces $S(\text{no } I)$ which do not enclose current sources.

After j_n has been obtained from equations (3.9.29) and (3.9.30), the potential can be calculated at any points of the model from equation (3.9.27).

The analytical form of the Green's function defined by equations (3.9.22) and (3.9.23) is known only for very simple space structures, such as the whole-space and the half-space. The whole-space Green's function $\mathscr{G}_0$ is obtained by solving equation (3.9.22) subject to the regularity condition at infinity:

$$\mathscr{G}_0(\mathbf{R}, \mathbf{R}_0) = \frac{1}{4\pi \left[(\tilde{\rho}/\rho)(\mathbf{R} - \mathbf{R}_0) \cdot (\mathbf{R} - \mathbf{R}_0) \right]^{1/2}} \qquad (3.9.31)$$

Expressed in a system of coordinates whose axes x, y, z coincide with the principal axes of anisotropy a, b, c, $\mathscr{G}_0$ acquires the simple form:

$$\mathscr{G}_0(\mathbf{R}, \mathbf{R}_0) = \frac{1}{4\pi \left[a(x - x_0)^2 + \beta(y - y_0)^2 + \gamma(z - z_0)^2 \right]^{1/2}} \qquad (3.9.32)$$

where $\mathbf{R} = (x, y, z)$ and $\mathbf{R}_0 = (x_0, y_0, z_0)$. The coefficients a, β, and γ are defined by equations (3.9.11) and (3.9.12). The half-space Green's function, in which one of the principal axes of resistivity is parallel to the ground surface, is given in Appendix C.

Let us finally examine the relation of the fictitious surface charge density

$$\sigma_{S\,\text{fict}} = \varepsilon_0 \rho \, j_n \qquad (3.9.33)$$

in equation (3.9.29) to the true surface charge density σ_S distributed on S. The true surface charge density is given by equation (2.5.5) as $\sigma_S = \varepsilon_0 (E_{2n} - E_{1n})$, where E_n is the normal component of the electric field on S. Since the potential U in V_1 is constant, the electric field in V_1 is zero, and in V_2 on S normal to S, that is, $E_{1n} = 0$, $\mathbf{E}_2 = \mathbf{E}_{2n}$. The true surface charge density on S is thus:

$$\sigma_S = \varepsilon_0 E_{2n} = \varepsilon_0 |\tilde{\rho}\, \mathbf{j}| \qquad (3.9.34)$$

Comparing equations (3.9.33) and (3.9.34), we can see that the fictitious surface charge density, which is the source function in equation (3.9.29), differs from the true charge density generated by current **j** on S. The difference is due to the gap made by conductor V_1 in the volume charge distribution resulting from the anisotropy of conductivity in region V_2.

3.9.4 Integral equation for an anisotropic body in an anisotropic environment

In the previous section we derived an integral equation solution for the boundary-value problem of potential in an anisotropic space containing a perfectly conductive anomalous body. Next we shall consider, after [40], a more general model in which both the anomalous body and the environment are electrically anisotropic.

The earth model is shown in Fig. 3.19, in which S is the interface between the anomalous body V_1 and its environment, and A represents the boundaries of the space surrounding the body. The primary current I is assumed to be emitted by a point electrode located at point $\mathbf{R}_P$ in region V_2. We shall attempt to find integral equation solutions for the potential functions u_1 and u_2, in which the source functions μ_1 and μ_2 in regions V_1 and V_2 are the fictitious normal surface current densities on either side of S. We express the potential in the form

$$U_1(\mathbf{R}) = \rho_1 \int_S \mathscr{G}_1(\mathbf{R}, \mathbf{R}_0) \mu_1(\mathbf{R}_0) \, dS_0 \qquad (3.9.35)$$

in region V_1, and

$$U_2(\mathbf{R}) = U_P(\mathbf{R}) + \rho_2 \int_S \mathscr{G}_2(\mathbf{R}, \mathbf{R}_0) \mu_2(\mathbf{R}_0) \, dS_0 \qquad (3.9.36)$$

in region V_2. In equations (3.9.35) and (3.9.36), $\tilde{\rho} = \tilde{g}^{-1}$, $\rho_i = (\rho_{ai} \rho_{bi} \rho_{ci})^{\frac{1}{3}}$, and ρ_{ai}, ρ_{bi} and ρ_{ci}, are the principal values of resistivity in regions V_i, $i = 1, 2$. The primary potential in (3.9.36) can be written in the form

$$U_P(\mathbf{R}) = \rho_2 \mathscr{G}_2(\mathbf{R}, \mathbf{R}_P) I(\mathbf{R}_P) \qquad (3.9.37)$$

The Green's functions of equations (3.9.35) and (3.9.36) satisfy the generalized Poisson's equation:

$$\nabla \cdot (\tilde{g}_i \nabla \mathscr{G}_i) = - g_i \delta(\mathbf{R} - \mathbf{R}_0) \qquad i = 1, 2 \qquad (3.9.38)$$

Green's function $\mathscr{G}_1$ for region V_1 satisfies the regularity condition at infinity but not the boundary conditions on surface A. This is because $\mathscr{G}_1$ is the whole-space Green's function which can be written in the form

$$\mathcal{G}_1(\mathbf{R}, \mathbf{R}_0) = \frac{1}{4\pi \left[(\tilde{\rho}_1/\rho_1)(\mathbf{R} - \mathbf{R}_0) \cdot (\mathbf{R} - \mathbf{R}_0)\right]^{1/2}} \quad (3.9.39)$$

Green's function $\mathcal{G}_2$ for region V_2 satisfies the boundary conditions

$$\mathcal{G}_2 = \mathcal{G}_3$$

$$(\tilde{g}_2 \nabla \mathcal{G}_2) \cdot \mathbf{n} = (\tilde{g}_3 \nabla \mathcal{G}_3) \cdot \mathbf{n} \quad (3.9.40)$$

on surface A, and the regularity condition at infinity, which yields $\mathcal{G}_2$ in the form

$$\mathcal{G}_2 = \mathcal{G}_{02} + \mathcal{G}_{S2} \quad (3.9.41)$$

The first term in equation (3.9.41) is the singular part of $\mathcal{G}_2$, which is obtained as the particular solution of equation (3.9.38), in the following form:

$$\mathcal{G}_{02}(\mathbf{R}, \mathbf{R}_0) = \frac{1}{4\pi \left[(\tilde{\rho}_2/\rho_2)(\mathbf{R} - \mathbf{R}_0) \cdot (\mathbf{R} - \mathbf{R}_0)\right]^{1/2}} \quad (3.9.42)$$

The second term $\mathcal{G}_{S2}$ is a non-singular function which represents the potential of the surface and volume charges caused by boundary A. The half-space Green's function for a model in which one of the principal axes is parallel to the ground surface A, is given in Appendix C.

To obtain suitable equations for solving the unknown source densities μ_1 and μ_2, we require boundary conditions such that the potential and the normal component of current density are continuous on S, that is,

$$U_1 = U_2 \quad (3.9.43a)$$

$$(\tilde{g}_1 \nabla U_1) \cdot \mathbf{n} = (\tilde{g}_2 \nabla U_2) \cdot \mathbf{n} \quad (3.9.43b)$$

on surface S.

In substituting the boundary conditions (3.9.43) into equations (3.9.35) and (3.9.36), the calculation points $\mathbf{R}$ in regions V_1 and V_2 are moved to the surface S. Boundary condition (3.9.43a) is satisfied by directly equating the potentials U_1 and U_2 represented by equations (3.9.35) and (3.9.36). The integrals in these equations are convergent on S and thus the singular points $\mathbf{R} = \mathbf{R}_0$ do not generate additional terms in the potential. In applying boundary condition (3.9.43b), the potentials U_1 and U_2 are differentiated so that the current density functions can be calculated. Green's function $\nabla \mathcal{G}_1$ of the integral representation obtained from equation (3.9.35) and the singular part $\nabla \mathcal{G}_{02}$ of Green's function $\nabla \mathcal{G}_2$, associated with the integral representation obtained from equation (3.9.36), each produce an additional term in dealing with the singular points $\mathbf{R} = \mathbf{R}_0$ of the integrals.

The singular Green's function for the field strength, which is needed for calculating the singular part of current density, is of the form

$$-\nabla \mathscr{G}_0(\mathbf{R}, \mathbf{R}_0) = \frac{(\tilde{\rho}/\rho)(\mathbf{R} - \mathbf{R}_0)}{4\pi\,[(\tilde{\rho}/\rho)(\mathbf{R} - \mathbf{R}_0) \cdot (\mathbf{R} - \mathbf{R}_0)]^{3/2}} \quad (3.9.44)$$

which, by applying the integral representation for the secondary current density

$$\mathbf{j}_{\text{sec}}(\mathbf{R}) = \tilde{g}\,\mathbf{E}(\mathbf{R}) = \rho \int_S -\tilde{g}\,\nabla\mathscr{G}(\mathbf{R},\mathbf{R}_0)\,\mu(\mathbf{R}_0)\,dS_0 \quad (3.9.45)$$

yields for the singular part of current density normal to surface S:

$$\mathbf{j}_{n\,\text{sing}}(\mathbf{R}) = \int_S \frac{(\mathbf{R}-\mathbf{R}_0)\cdot\mathbf{n}\,\mu(\mathbf{R}_0)\,dS_0}{4\pi\,[(\tilde{\rho}/\rho)(\mathbf{R}-\mathbf{R}_0)\cdot(\mathbf{R}-\mathbf{R}_0)]^{3/2}} \quad (3.9.46)$$

In the limit, as $\mathbf{R}$ approaches surface S from either side of S, equation (3.9.46) yields the current densities $j_{n1}(\mathbf{R}=\mathbf{R}_0) = -\mu_1/2$ in V_1, and $j_{n2}(\mathbf{R}=\mathbf{R}_0) = \mu_2/2$ in V_2.

We can now write the pair of integral equations for solving the source densities μ_1 and μ_2 on S:

$$\rho_1 \int_S \mathscr{G}_1(\mathbf{R},\mathbf{R}_0)\,\mu_1(\mathbf{R}_0)\,dS_0 = U_\text{P}(\mathbf{R}) + \rho_2 \int_S \mathscr{G}_2(\mathbf{R},\mathbf{R}_0)\,\mu_2(\mathbf{R}_0)\,dS_0 \quad (3.9.47)$$

$$-\frac{\mu_1}{2} + \rho_1 \int_S -[\tilde{g}_1\,\nabla\mathscr{G}_1(\mathbf{R},\mathbf{R}_0)]\cdot\mathbf{n}\,\mu_1(\mathbf{R}_0)\,dS_0 = -[\tilde{g}_2\nabla U_\text{P}(\mathbf{R})]\cdot\mathbf{n} + \frac{\mu_2}{2}$$
$$+ \rho_2 \int_S -[\tilde{g}_2\nabla\mathscr{G}_2(\mathbf{R},\mathbf{R}_0)]\cdot\mathbf{n}\,\mu_2(\mathbf{R}_0)\,dS_0 \quad (3.9.48)$$

where $\mathbf{R}$ and $\mathbf{R}_0$ are located on surface S. In equation (3.9.47) the integrals are convergent, and in equation (3.9.48) they are defined as the principal values.

After the fictitious surface charge functions μ_1 and μ_2 have been solved from the set of equations (3.9.47) and (3.9.48), the potential can be calculated at any arbitrary point of region V_1 from equation (3.9.35), and of region V_2 from equation (3.9.36).

4

Elements of magnetostatics

4.1 INTRODUCTION

A fundamental feature of Maxwell's equations is their symmetry in relation to the electric and magnetic fields. Therefore, the static magnetic field problem, in the absence of currents, is formally analogous to the electrical field problem. Owing to this analogy, the sources of anomalous magnetic fields can be represented by magnetic pole distributions. Since physically real magnetic monopoles have not been found, they should be regarded merely as equivalent sources.

The true physical character of the sources of magnetic anomalies are in fact variations in density and orientation of atomic currents. Each atomic loop can be uniquely represented by a magnetic dipole located at the centre of the loop and orientated in the direction of the loop axis. This makes it possible to represent the sources of the magnetic field by volume distributions of magnetic dipoles. The formulation of magnetic anomalies by using volume dipole distributions as sources is common practice in applied geophysics.

A unique equivalence exists between the formulations based on polar and dipolar sources. They are simply related by the divergence theorem of vector analysis.

4.2 INTEGRAL REPRESENTATION OF MAGNETIC POTENTIAL

We shall first define the boundary-value problem for the magnetic scalar potential, and then give the principles by which this boundary-value problem can be converted into an integral equation.

The constitutive relation between the magnetic field **H** and flux density **B** can be represented, by analogy with Ohm's law, in the following form:

$$\mathbf{H} = \frac{1}{\mu}\mathbf{B} \qquad (4.2.1)$$

where μ [V s A^{-1} m^{-1}] is the absolute permeability of the medium. In this book, permeability μ is considered to be linear and isotropic.

From the vector identity

$$\nabla \times \nabla U^* = 0 \qquad (4.2.2)$$

it follows that if the curl of the magnetic field vanishes, then the field can be expressed as the gradient of a scalar potential. For a time-independent system, we obtain from Maxwell's equation (2.1.2) the expression:

$$\nabla \times \mathbf{H} = \mathbf{j} \qquad (4.2.3)$$

where $\mathbf{j}$ is current density. In the absence of currents, $\mathbf{H}$ is thus purely laminar and can be expressed as follows:

$$\mathbf{H} = -\nabla U^* \qquad (4.2.4)$$

Taking the divergence on both sides of equation (4.2.4), we obtain, referring to Fig. 4.1, for magnetic scalar potential U^* Poisson's equation

$$\nabla^2 U^* = -\sigma^* \qquad (4.2.5)$$

in the anomalous region V_1, where the volume density of magnetic poles, σ^* [A m^{-2}], is given by Gauss's law for the magnetic field:

$$\nabla \cdot \mathbf{H} = \sigma^* \qquad (4.2.6)$$

In region V_2, which is free of magnetic poles, the potential satisfies Laplace's equation

$$\nabla^2 U^* = 0 \qquad (4.2.7)$$

Since Maxwell's equations imply that the tangential component of the magnetic field and the normal component of magnetic flux density are continuous when passing through physical boundaries, we require, according to Fig. 4.1, that potential U^* satisfies the boundary conditions

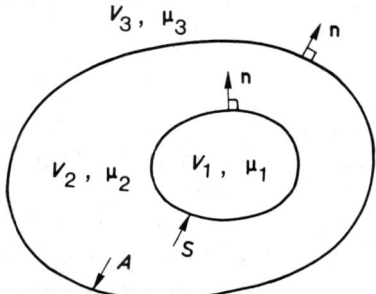

Figure 4.1 Earth model for magnetic potential problem.

$$U_3^* = U_2^* \tag{4.2.8a}$$

$$\mu_3 \frac{\partial U_3^*}{\partial n} = \mu_2 \frac{\partial U_2^*}{\partial n} \tag{4.2.8b}$$

on the surface A.

We shall now convert the magnetostatic problem defined by equations (4.2.5), (4.2.7) and (4.2.8) into an integral equation by using the Green's function technique, in the same way as was done in Chapter 2 for an electrical problem.

The Green's function of the integral equation to be derived is taken to satisfy Poisson's equation

$$\nabla^2 G = -\delta(\mathbf{R} - \mathbf{R}_0) \tag{4.2.9}$$

and the same boundary conditions as the potential, namely:

$$G_3 = G_2 \tag{4.2.10a}$$

$$\mu_3 \frac{\partial G_3}{\partial n} = \mu_2 \frac{\partial G_2}{\partial n} \tag{4.2.10b}$$

As the solution of equations (4.2.9) and (4.2.10), Green's function for the magnetic scalar potential can be written in the following form:

$$G = G_0 + G_S \tag{4.2.11}$$

G_0 is the singular particular solution of equation (4.2.9), known as the **whole-space Green's function**. It is regular at infinity, and, together with its derivative, automatically continuous on surface A. Under these circumstances we obtain for G_0 the following expression:

$$G_0 = \frac{1}{4\pi |\mathbf{R} - \mathbf{R}_0|} \tag{4.2.12}$$

The second part, G_S, of Green's function G is a non-singular function which represents the potential generated by the pole distributions accumulated on boundaries A of the space surrounding the anomalous body V_1 (see Fig. 4.1). G_S is the solution of Laplace's equation:

$$\nabla^2 G_S = 0 \tag{4.2.13}$$

subject to the boundary conditions, given by equations (4.2.10), on surface A.

In magnetic modelling problems, the surrounding space is usually considered to be an empty space, for which $G = G_0$.

Consider region $V_2 + V_1$ of Fig. 4.1, where all the anomalous pole distributions are located in region V_1. The physical boundaries of the space $V_2 + V_3$, surrounding the anomalous region V_1, are represented by surface A. Let us substitute functions U^* and G into Green's second identity and apply the identity to the region $V_2 + V_1$:

$$\int_{V_1+V_2} (G_2 \nabla^2 U_2^* - U_2^* \nabla^2 G_2) \, dV = \int_A \left(G_2 \frac{\partial U_2^*}{\partial n} - U_2^* \frac{\partial G_2}{\partial n} \right) dA \quad (4.2.14)$$

where the differentiations and integrations are carried out with respect to variable **R**, while the point $\mathbf{R}_0$ is kept fixed in region V_2. Substituting equations (4.2.5) and (4.2.9), equation (4.2.14) acquires the form

$$U^* = \int_{V_1} G_2 \sigma^* dV + \int_A \left(G_2 \frac{\partial U_2^*}{\partial n} - U_2^* \frac{\partial G_2}{\partial n} \right) dA \quad (4.2.15)$$

We next apply Green's second identity to region V_3 while keeping point $\mathbf{R}_0$ in region V_2. Since U^* and G are regular at infinity, the integral over the outer boundary of region V_3 vanishes at infinity, so that the identity simplifies to

$$0 = -\int_A \left(G_3 \frac{\partial U_3^*}{\partial n} - U_3^* \frac{\partial G_3}{\partial n} \right) dA \quad (4.2.16)$$

Multiplying equation (4.2.16) by μ_2/μ_3, adding the respective sides of this equation and equation (4.2.15), and substituting boundary conditions (4.2.8) and (4.2.10), we obtain for U^*

$$U^* = \int_{V_1} G_2 \sigma^* dV \quad (4.2.17)$$

which, after interchanging the variables **R** and $\mathbf{R}_0$ and using the reciprocity property of Green's function, yields:

$$U^*(\mathbf{R}) = \int_{V_1} G(\mathbf{R}, \mathbf{R}_0) \, \sigma^*(\mathbf{R}_0) dV_0 \quad (4.2.18)$$

The following three sections deal with the properties of the commonly encountered source distributions of magnetic scalar potential: volume and surface distributions of simple magnetic poles, and the volume distribution of magnetic dipoles.

4.3 VOLUME DISTRIBUTION OF POLES

The volume density of magnetic poles σ^* can be represented in terms of the magnetic field **H** with the aid of Gauss's law for the magnetic field

$$\nabla \cdot \mathbf{H} = \sigma^* \quad (4.3.1)$$

Substituting the constitutive relation $\mathbf{B} = \mu \mathbf{H}$ into equation (4.3.1), and applying Maxwell's equation

$$\nabla \cdot \mathbf{B} = 0 \quad (4.3.2)$$

98 Elements of magnetostatics

we obtain for the volume density of magnetic poles:

$$\sigma^* = \mathbf{B} \cdot \nabla\left(\frac{1}{\mu}\right) \tag{4.3.3}$$

The potential at the point $\mathbf{R}$ outside the source region V_1, in which there is a pole density σ^* at point $\mathbf{R}_0$, is represented by the integral

$$U^*(\mathbf{R}) = \int_{V_1} G(\mathbf{R}, \mathbf{R}_0)\, \sigma^*(\mathbf{R}_0)\,dV_0 \tag{4.3.4}$$

When the calculation point $\mathbf{R}$ is in source region V_1, the integral in equation (4.3.4) is defined as the limit

$$U^*(\mathbf{R}) = \lim_{v \to 0} \int_{V_1 - v} G(\mathbf{R}, \mathbf{R}_0)\, \sigma^*(\mathbf{R}_0)\,dV_0 \tag{4.3.5}$$

where v is a small volume containing point $\mathbf{R}$. The integral is improper but the contribution of volume v vanishes as v approaches zero, and the integral is convergent and representable by equation (4.3.4).

By a similar argument it can be shown that, for an integral representation of the field strength, the contribution of volume v vanishes as v approaches zero, and the integral thus converges to the following form:

$$\mathbf{H}(\mathbf{R}) = -\int_{V_1} \nabla G(\mathbf{R}, \mathbf{R}_0)\, \sigma^*(\mathbf{R}_0)\,dV_0 \tag{4.3.6}$$

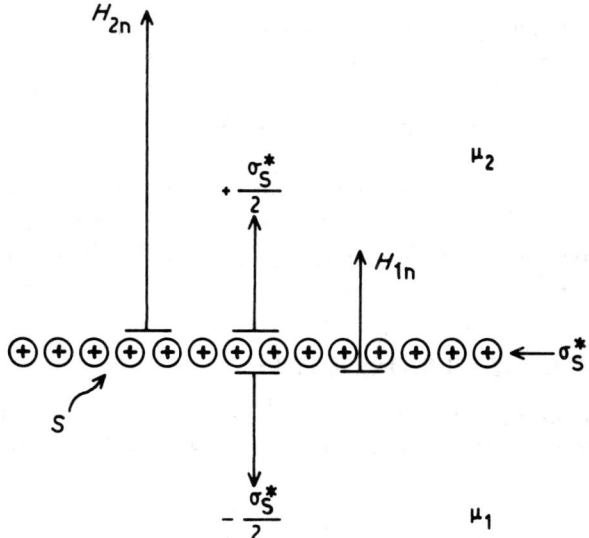

Figure 4.2 Behaviour of normal magnetic field due to surface pole distribution on a permeability discontinuity.

4.4 SURFACE DISTRIBUTION OF POLES

Discontinuity surfaces of magnetic permeability are the accumulation regions of surface poles. Consider a model as illustrated in Fig. 4.2, where S is the boundary between regions 1 and 2 having permeabilities μ_1 and μ_2. The magnetic scalar potential U^* at point $\mathbf{R}$, produced by the poles distributed on the surface S with density $\sigma_S^*(\mathbf{R}_0)$ [A m^{-1}], is

$$U^*(\mathbf{R}) = \int_S G(\mathbf{R}, \mathbf{R}_0)\, \sigma_S^*(\mathbf{R}_0)\, dS_0 \qquad (4.4.1)$$

When $\mathbf{R}$ is on surface S, the potential is defined as the limit

$$U^*(\mathbf{R}) = \lim_{s \to 0} \int_{S-s} G(\mathbf{R}, \mathbf{R}_0)\, \sigma_S^*(\mathbf{R}_0)\, dS_0 \qquad (4.4.2)$$

where s is a small surface containing the calculation point $\mathbf{R}$. The contribution of the surface charge on s vanishes as s approaches zero, and the integral is convergent and representable by equation (4.4.1).

The field strength $\mathbf{H}$ outside the surface S can be represented as follows:

$$\mathbf{H}(\mathbf{R}) = -\int_S \nabla G(\mathbf{R}, \mathbf{R}_0)\, \sigma_S^*(\mathbf{R}_0)\, dS_0 \qquad (4.4.3)$$

where the differentiation of Green's function is performed with respect to variable $\mathbf{R}$.

The limits of the normal component of the field at points $\mathbf{R}_1$ and $\mathbf{R}_2$ of regions 1 and 2, as $\mathbf{R}_1$ and $\mathbf{R}_2$ approach $\mathbf{R}$ on S along the common normal $\mathbf{n}$, are as follows:

$$\lim_{\mathbf{R}_2 \to \mathbf{R}} H_n(\mathbf{R}_2) = \frac{\sigma_S^*(\mathbf{R})}{2} - \int_S \frac{\partial G(\mathbf{R}, \mathbf{R}_0)}{\partial n}\, \sigma_S^*(\mathbf{R}_0)\, dS_0 \qquad (4.4.4a)$$

$$\lim_{\mathbf{R}_1 \to \mathbf{R}} H_n(\mathbf{R}_1) = -\frac{\sigma_S^*(\mathbf{R})}{2} - \int_S \frac{\partial G(\mathbf{R}, \mathbf{R}_0)}{\partial n}\, \sigma_S^*(\mathbf{R}_0)\, dS_0 \qquad (4.4.4b)$$

where the surface integrals are improper convergent integrals.

The boundary condition satisfied by the normal component of field strength on S can now be obtained from equations (4.4.4) as follows:

$$H_{2n} - H_{1n} = \sigma_S^* \qquad (4.4.5)$$

which can be considered to represent Gauss's law for a surface distribution of poles.

Maxwell's equation (2.1.3) implies that the normal component of magnetic flux density $\mathbf{B}$ is continuous on S:

$$B_{1n} = B_{2n} \qquad (4.4.6)$$

Applying the constitutive relation $\mathbf{B} = \mu \mathbf{H}$ and substituting equation (4.4.6) into equation (4.4.5), we obtain for the surface pole density:

$$\sigma_S^* = \left(\frac{1}{\mu_2} - \frac{1}{\mu_1} \right) B_n \qquad (4.4.7)$$

which is the relation between the surface density of poles and the change in permeability across boundary S.

4.5 VOLUME DISTRIBUTION OF DIPOLES

We shall now give an integral representation for potential U^* produced by a volume distribution of magnetic dipoles. This representation is physically appropriate because the atomic currents, which are the true sources of magnetic anomalies, can be uniquely represented by equivalent magnetic dipoles.

We start by considering the sum of the potentials due to volume pole distribution in V_1 and surface pole distribution on S, which can be written according to equations (4.3.4) and (4.4.1) as follows:

$$U^*(\mathbf{R}) = \int_{V_1} G(\mathbf{R}, \mathbf{R}_0) \sigma^*(\mathbf{R}_0) dV_0 + \int_S G(\mathbf{R}, \mathbf{R}_0) \sigma_S^*(\mathbf{R}_0) dS_0 \qquad (4.5.1)$$

where $\sigma^* = \nabla \cdot \mathbf{H}$ and $\sigma_S^* = H_{2n} - H_{1n}$. To substitute the dipole moment density for field strength $\mathbf{H}$, use is made of the relation

$$\mathbf{H} = \frac{1}{\mu_0} \mathbf{B} - \mathbf{M} \qquad (4.5.2)$$

where $\mathbf{M}\,[A\,m^{-1}]$ is the magnetic dipole moment per unit volume, known as the **intensity of magnetization**. Using Maxwell's equation $\nabla \cdot \mathbf{B} = 0$, we obtain from equation (4.5.2)

$$\nabla \cdot \mathbf{H} = - \nabla \cdot \mathbf{M} \qquad (4.5.3)$$

Correspondingly, using the boundary condition $B_{2n} = B_{1n}$ and equation (4.5.2), yields

$$H_{2n} - H_{1n} = - (M_{2n} - M_{1n}) \qquad (4.5.4)$$

Assume, as is usual for interpretation problems in applied geophysics, that the magnetization M_2 in region V_2 is zero. With the above substitutions, equation (4.5.1) acquires the following form:

$$U^* = \int_{V_1} - G \nabla_0 \cdot \mathbf{M}_1 \, dV_0 + \int_S M_{1n} \, G \, dS_0 \qquad (4.5.5)$$

Dropping subscript 1 and using the partial differentiation rule

$$G \nabla \cdot \mathbf{M} = \nabla \cdot (G\mathbf{M}) - \mathbf{M} \cdot \nabla G$$

equation (4.5.5) transforms into

$$U^* = \int_{V_1} -\nabla_0 \cdot (G\mathbf{M}) \, dV_0 + \int_{V_1} \mathbf{M} \cdot \nabla_0 G \, dV_0 + \int_S M_n G \, dS_0 \qquad (4.5.6)$$

Using the divergence theorem, we finally obtain for the magnetic potential:

$$U^*(\mathbf{R}) = \int_{V_1} \mathbf{M}(\mathbf{R}_0) \cdot \nabla_0 G(\mathbf{R}, \mathbf{R}_0) dV_0 \qquad (4.5.7)$$

where the gradient is taken with respect to $\mathbf{R}_0$.

Equation (4.5.7) is vectorial, that is, three components of magnetization $\mathbf{M}$ must be solved in the source region. Furthermore, the singularity of Green's function for dipoles is of a higher order than that for simple poles. Therefore, equation (4.5.7) is not very practical as a basis for integral equation formulation. Nevertheless, it is commonly applied, by using rough but reasonable approximations, to calculate magnetic anomalies for models having low susceptibility values.

5

Magnetic methods

5.1 MAGNETIC PROPERTIES OF ROCKS

Magnetic anomalies in bedrock are practically entirely governed by the magnetic properties, mode of distribution and amount of ferrimagnetic minerals. The most common of these is **magnetite** (Fe_3O_4), but **pyrrhotite** (FeS) may also cause magnetic anomalies.

Ferrimagnetic crystals consist of numerous domains. The electron-spin vectors of neighbouring atoms are aligned within the domains but not at the boundaries between domains. Consequently the resultant magnetic moments of different domains are normally directed differently but, when subjected to a magnetizing field, are susceptible to polarization. Parallel alignment is opposed by thermal agitation so that, above a certain temperature called the **Curie point**, the orientation of the moments is random. For magnetite, the Curie point is 580°C.

The total magnetization of ferrimagnetic rocks is the resultant of induced and remanent magnetization. The **induced** magnetization is determined by the earth's magnetic field and by the present mode of distribution and chemical composition of magnetic minerals. The natural **remanent** magnetization is generated by geological processes during the rock's history.

The induced magnetization M_i in the rocks is controlled by the susceptibility k, defined by $M_i = k H$ where H is the magnetic field strength. For rocks containing magnetite, susceptibility values have been observed to vary from 25×10^{-6} SI for granite having a weak dissemination of magnetite, up to 15 SI for pure magnetite [9]. The ratio of remanent magnetization M_r to the induced magnetization M_i is represented by the Koenigsberger ratio $Q = M_r/M_i$. Q values from 0.3–10 for granites and up to 100–118 for basalts have been measured [9].

For low to moderate external magnetic field strengths, k is constant. For high field strengths, the susceptibility decreases with increasing field strength while the magnetization of the specimen approaches saturation. This non-linear behaviour of susceptibility is comparable with that of resistivity. The earth's magnetic field is so weak, however, that the models treated in magnetic modelling by applied geophysicists can be considered to be linear, that is, the susceptibility is independent of the magnetizing field strength.

5.2 HIGH-SUSCEPTIBILITY MODELS

In most geological formations, susceptibility values are so low that the integral equations for the source densities producing their anomalies can be solved by using various approximation techniques. The demagnetization factor technique can be applied with sufficient accuracy as long as the susceptibility of the anomalous body does not exceed 1 SI [41, 42]. Above this limit, more sophisticated solution methods must be used. Consequently, such important geological formations as magnetite ores are beyond the range of validity of these approximation techniques.

In the following we shall present integral equation solutions for magnetic anomalies caused by earth models whose susceptibility values exceed 1 SI. The models are assumed to possess only induced magnetization; the contribution of remanent magnetization is considered in Section 5.5. The models under consideration are composed of regions having homogeneous susceptibilities, in which case the field problems can be represented by surface integral equations.

5.2.1 Three-dimensional models

Consider an earth model, illustrated by Fig. 4.1, in which anomalous body V_1 with homogeneous permeability μ_1 is bounded by region V_2 with permeability μ_2. The space surrounding V_1 may also contain permeability boundaries, represented in Fig. 4.1 by surface A. The sources of the primary field are assumed to be located outside the model.

The magnetic scalar potential in the model can be written in the form

$$U^*(\mathbf{R}) = U_P^*(\mathbf{R}) + \int_S G(\mathbf{R}, \mathbf{R}_0) \sigma_S^*(\mathbf{R}_0) dS_0 \qquad (5.2.1)$$

where U_P^* is the primary potential and σ_S^* is the unknown magnetic surface pole density on surface S. σ_S^* represents the physical sources for the magnetic anomaly of body V_1. Green's function G is defined in equations (4.2.9)–(4.2.13).

To obtain the expression for field strength $\mathbf{H}$, we use the relation $\mathbf{H} = -\nabla U^*$ so that

$$\mathbf{H}(\mathbf{R}) = \mathbf{H}_P(\mathbf{R}) - \int_S \nabla G(\mathbf{R}, \mathbf{R}_0) \sigma_S^*(\mathbf{R}_0) dS_0 \qquad (5.2.2)$$

which is the basic integral representation for the field problem under consideration.

To form an integral equation for determining σ_S^*, we move the calculation point **R** to surface S in the region V_2 exterior to S. (The same result would be obtained by considering the region V_1 interior to S). Replacing the singular point $\mathbf{R} = \mathbf{R}_0$ from the surface integral in accordance with equation (4.4.4a), neglecting remanence, and using equations (4.4.5)–(4.4.7), we obtain the integral equation

$$\frac{\mu_1 + \mu_2}{2(\mu_1 - \mu_2)} \sigma_S^*(\mathbf{R}) = H_{\mathrm{Pn}}(\mathbf{R}) - \int_S \frac{\partial G(\mathbf{R}, \mathbf{R}_0)}{\partial n} \sigma_S^*(\mathbf{R}_0) \mathrm{d} S_0 \qquad (5.2.3)$$

where the surface integral is defined in the sense of the principal value. If the constitutive relation between **H** and **B** is given in terms of susceptibility k, it may be practical to substitute $\mu = \mu_0 (1 + k)$ in equation (5.2.3), and the coefficient of σ_S^* on the left-hand side of the equation is then transformed to

$$\frac{\mu_1 + \mu_2}{2(\mu_1 - \mu_2)} = \frac{2 + k_1 + k_2}{2(k_1 - k_2)} \qquad (5.2.4)$$

After numerical solution of equation (5.2.3) for pole density σ_S^* on surface S, the field strength **H** can be calculated at any arbitrary point of regions V_1, V_2 and V_3 by using equation (5.2.2).

In geophysical applications, the theoretical model can in most cases be represented by a magnetized body located in a non-magnetized environment. In Fig. 4.1, then, $\mu_2 = \mu_3 = \mu_0$. For these models, Green's function is the whole-space Green's function, $G = G_0$, as given by equation (4.2.12).

Green's function for a magnetized half-space can be obtained by requiring G to satisfy the boundary conditions on ground surface A given by equation (4.2.10). Assuming that A is represented by the plane $z = 0$ and substituting $\mu_3 = \mu_0$ in equation (4.2.8), we obtain the following half-space Green's functions G_2 and G_3 for regions V_2 and V_3:

$$G_2 = \frac{1}{4\pi} \left\{ \frac{1}{|\mathbf{R} - \mathbf{R}_0|} + \frac{K}{|\mathbf{R} - \mathbf{R}_0'|} \right\} \qquad (5.2.5)$$

$$G_3 = \frac{1 + K}{4\pi |\mathbf{R} - \mathbf{R}_0|} \qquad (5.2.6)$$

where $\mathbf{R}_0 = (x_0, y_0, z_0)$, $\mathbf{R}_0' = (x_0, y_0, -z_0)$, and $K = (\mu_2 - \mu_0)/(\mu_2 + \mu_0)$.

For the magnetized half-space environment, the primary fields above and below the ground surface differ from each other. Assume that the vertical components of the homogeneous primary fields $\mathbf{H}_{\mathrm{P2}}$ and $\mathbf{H}_{\mathrm{P3}}$ in regions V_2 and V_3 are H_{Pz2} and H_{Pz3}, and the corresponding horizontal components are H_{Ph2} and H_{Ph3}. Taking the tangential field to be continuous across boundary A, and using boundary condition (4.2.8b), the field components in region V_2 are related to the corresponding components in region V_3 by the formulas:

$$\mathbf{H}_{\text{P}z2} = \frac{\mu_0}{\mu_2} \mathbf{H}_{\text{P}z3}$$

$$\mathbf{H}_{\text{Ph}2} = \mathbf{H}_{\text{Ph}3} \qquad (5.2.7)$$

In magnetic modelling, the magnetized half-space is not such a widely applicable host structure as the conductive half-space counterpart in electrical modelling. This is because the magnetic primary field, due to its homogeneity, does not suffer geometrical attenuation and, hence, boundary conditions (4.2.8) should be satisfied both on the ground surface and on other large discontinuity surfaces of permeability, such as the Curie isotherm surface. If the magnetic formation to be modelled requires the magnetization of the region surrounding a three-dimensional anomalous body to be taken into account, then it can be simulated most appropriately by a larger three-dimensional magnetized body enclosing the anomalous body.

In the following, we shall give an integral equation solution for the magnetic field problem in the model represented by Fig. 4.1. Only induced magnetization is taken into account, the effect of remanence being considered in Section 5.5. Surface S is the boundary between region V_1, of permeability μ_1 and region V_2 of permeability μ_2, and surface A is the boundary between V_2 and the whole-space V_3 external to it, of permeability μ_0. The field strength in the model can be written as:

$$\mathbf{H}(\mathbf{R}) = \mathbf{H}_\text{P} - \int_A \nabla G(\mathbf{R}, \mathbf{R}_0) \sigma_S^*(\mathbf{R}_0) \mathrm{d}A_0 - \int_S \nabla G(\mathbf{R}, \mathbf{R}_0) \sigma_S^*(\mathbf{R}_0) \mathrm{d}S_0 \qquad (5.2.8)$$

where $\mathbf{H}_\text{P}$ is the homogeneous primary field generated by distant sources in region V_3.

To obtain the pole density σ_S^* on surfaces A and S, we transform equation (5.2.8) into an integral equation, first by moving the calculation point $\mathbf{R}$ in region V_3 to surface A, and then from region V_2 to surface S. Replacing the contributions of singular points $\mathbf{R} = \mathbf{R}_0$ from the surface integrals over A and S according to equation (4.4.4a) and substituting equations (4.4.5)–(4.4.7), we obtain:

$$\frac{\mu_2 + \mu_3}{2(\mu_2 - \mu_3)} \sigma_S^*(\mathbf{R}) = H_{\text{P}n}(\mathbf{R}) - \int_A \frac{\partial G(\mathbf{R}, \mathbf{R}_0)}{\partial n} \sigma_S^*(\mathbf{R}_0) \mathrm{d}A_0$$

$$- \int_S \frac{\partial G(\mathbf{R}, \mathbf{R}_0)}{\partial n} \sigma_S^*(\mathbf{R}_0) \mathrm{d}S_0 \qquad (5.2.9\text{a})$$

where $\mathbf{R}$ is on A and $\mathbf{R}_0$ is on A and S, and

$$\frac{\mu_1 + \mu_2}{2(\mu_1 - \mu_2)} \sigma_S^*(\mathbf{R}) = H_{\text{P}n}(\mathbf{R}) - \int_A \frac{\partial G(\mathbf{R}, \mathbf{R}_0)}{\partial n} \sigma_S^*(\mathbf{R}_0) \mathrm{d}A_0$$

$$- \int_S \frac{\partial G(\mathbf{R}, \mathbf{R}_0)}{\partial n} \sigma_S^*(\mathbf{R}_0) \mathrm{d}S_0 \qquad (5.2.9\text{b})$$

where **R** is on S and $\mathbf{R}_0$ is on A and S. The surface integral over A in equation (5.2.9a) and the surface integral over S in equation (5.2.9b) are defined as the principal values.

The pole density σ_S^* on surfaces A and S can be obtained by simultaneously solving equations (5.2.9a) and (5.2.9b). After σ_S^* is known, the magnetic field can be calculated at any point of the model by using equation (5.2.8).

5.2.2 $2\frac{1}{2}$-dimensional models

As in the case of conductive geological formations, magnetic formations commonly have one horizontal dimension considerably longer than the other two. The anomalies generated by these structures can be most practically calculated by using $2\frac{1}{2}$ dimensional models. Since the primary magnetic field can be considered as homogeneous, the $2\frac{1}{2}$ dimensional magnetic model is simply defined as a truncated two-dimensional model, in which the effects of the end faces are disregarded in the calculations.

Suppose that the long dimension of the body is orientated parallel to the y-axis, and that the end faces are located at $y = -L/2$ and $y = L/2$. The magnetic field component perpendicular to the y-axis can be expressed in the plane $y = 0$ as:

$$\mathbf{H}_r(\mathbf{r}) = \mathbf{H}_{\text{Pr}} - \int_{-L/2}^{L/2} \int_C \nabla_S G(\mathbf{r}, \mathbf{r}_0; y_0) \sigma_S^*(\mathbf{r}_0; y_0) dC_0 dy_0 \quad (5.2.10)$$

where C is the contour of the body in the plane $y = 0$, $\mathbf{R} = \mathbf{r} + \mathbf{y}$, $\mathbf{r} = \mathbf{x} + \mathbf{z}$ and ∇_S denotes the surface gradient operating in plane $y = 0$. Assuming that the surface pole density σ_S^* can be considered to be independent on coordinate y, equation (5.2.10) can be written in the form

$$\mathbf{H}_r(\mathbf{r}) = \mathbf{H}_{\text{Pr}} - \int_C \nabla_S \overline{G}(\mathbf{r}, \mathbf{r}_0) \sigma_S^*(\mathbf{r}_0) dC_0 \quad (5.2.11)$$

$$\overline{G}(\mathbf{r}, \mathbf{r}_0) = \int_{-L/2}^{L/2} G(\mathbf{r}, \mathbf{r}_0; y_0) dy_0 \quad (5.2.12)$$

The unknown $2\frac{1}{2}$-dimensional surface pole density is obtained by transforming equation (5.2.11) into an integral equation along curve C. By moving the calculation point $\mathbf{r}$ to C in the external region V_2 and treating the singular point $\mathbf{r} = \mathbf{r}_0$ in the line integral using a method analogous to that used in the preceding section for the case of a three-dimensional model, we obtain for σ_S^*:

$$\frac{\mu_1 + \mu_2}{2(\mu_1 - \mu_2)} \sigma_S^*(\mathbf{r}) = H_{\text{Pn}}(\mathbf{r}) - \int_C \frac{\partial \overline{G}(\mathbf{r}, \mathbf{r}_0)}{\partial n} \sigma_S^*(\mathbf{r}_0) dC_0 \quad (5.2.13)$$

where H_{Pn} is the component of the primary field normal to curve C and the y-axis.

Integral equation (5.2.13) can be solved for σ_s^* by standard numerical methods, after which the magnetic field component normal to the y-axis can be calculated at any point $\mathbf{r}$ of the plane $y = 0$ from equation (5.2.11).

The $2\frac{1}{2}$-dimensional Green's function for the whole-space can be obtained by substituting equation (4.2.12) into equation (5.2.12), so that

$$\overline{G}_0(\mathbf{r}, \mathbf{r}_0) = \frac{1}{4\pi} \ln \frac{(L/2) + [(L/2)^2 + |\mathbf{r} - \mathbf{r}_0|^2]^{\frac{1}{2}}}{-(L/2) + [(L/2)^2 + |\mathbf{r} - \mathbf{r}_0|^2]^{\frac{1}{2}}} \quad (5.2.14)$$

Substituting equation (5.2.5) and (5.2.6) into equation (5.2.12), the half-space Green's functions for regions V_3 and V_2 (see Fig. 4.1) can be expressed in the following form:

$$\overline{G}_3 = (1 + K)\overline{G}_0 \quad (5.2.15)$$

$$\overline{G}_2 = \overline{G}_0 + \frac{K}{4\pi} \ln \frac{(L/2) + [(L/2)^2 + |\mathbf{r} - \mathbf{r}_0'|^2]^{\frac{1}{2}}}{-(L/2) + [(L/2)^2 + |\mathbf{r} - \mathbf{r}_0'|^2]^{\frac{1}{2}}} \quad (5.2.16)$$

where $K = (\mu_2 - \mu_3)/(\mu_2 + \mu_3)$, $\mathbf{r}_0 = (x_0, y_0, z_0)$, $\mathbf{r}_0' = (x_0, y_0, -z_0)$ and surface A is represented by the plane $z = 0$.

In practice, application of the $2\frac{1}{2}$-dimensional model is advantageous when the strike length of the anomalous body is at least five times the next longest dimension of the body. In $2\frac{1}{2}$-dimensional modelling, the body is not divided into subsections along the direction of strike. Thus, rather complicated structural cross-sections can be modelled with considerable accuracy along the line of the body.

5.2.3 Thin sheet model

A thin magnetized sheet is a useful model in simulating magnetite ore veins or dikes. The permeability of the thin magnetized sheet must be higher than that of the environment, and its thickness is considered to be so small that the anomalous field is inseparably dependent upon the product of permeability and thickness of the sheet. We shall next solve the magnetic field problem for a thin sheet model by using integrodifferential as well as integral equations.

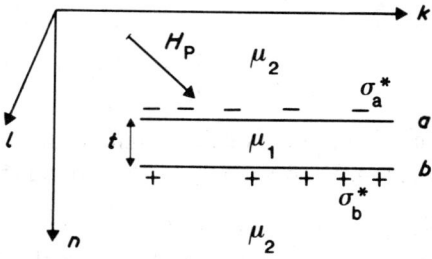

Figure 5.1 Part of a thin magnetized sheet. Permeability of the sheet μ_1 and that of the bounding medium μ_2. Coordinate system (k, l, n) fixed to the sheet.

108 *Magnetic methods*

The solution is closely analogous to that given in Section 3.3.6 for a thin conductor.

Consider a magnetized sheet of permeability μ_1 located in an environment of permeability μ_2 (Fig. 5.1). After introduction of the primary magnetic field $\mathbf{H}_P$, the model becomes magnetized and, as a result, a magnetic pole distribution of density σ_S^* is generated on surfaces S (the faces and the edges) of the sheet. In the solution of the field problem, the sheet is assumed to be so thin that its adjacent faces are joined, thus forming a single surface A bearing a net pole distribution of density σ_A^*. Letting σ^* denote the total pole density in the sheet, consisting of σ_A^* distributed on the faces, and of σ_C^* distributed on the thin edges C of the sheet, the magnetic scalar potential $U^*(\mathbf{R})$ in the model can be written as follows:

$$U^*(\mathbf{R}) = U_P^*(\mathbf{R}) + \int_A G(\mathbf{R}, \mathbf{R}_0) \sigma^*(\mathbf{R}_0) dA_0 \qquad (5.2.17)$$

where U_P^* is the potential of the primary field, G is Green's function for the space surrounding sheet A, and $\sigma^* = \sigma_A^* + \sigma_C^*$.

Let the adjacent faces of the sheet be located at $n = a$ and $n = b$, using the Cartesian system of coordinates (k, l, n), as shown in Fig. 5.1. Considering only the induced magnetization, the pole densities on surfaces $n = a$ and $n = b$ can be written in accordance with equation (4.4.7) as follows:

$$\sigma_a^* = \left(\frac{1}{\mu_1} - \frac{1}{\mu_2}\right) B_{na}$$

$$\sigma_b^* = \left(\frac{1}{\mu_2} - \frac{1}{\mu_1}\right) B_{nb} \qquad (5.2.18)$$

where B_{na} and B_{nb} are the n-components of the magnetic flux density on surfaces $n = a$ and $n = b$.

Let μ_1 be very high and the thickness, $t = b - a$, be very small in such a way that the product $g^* = \mu_1 t$ is finite. In the limit, the pole distributions σ_a^* and σ_b^* join, forming a simple layer of poles distributed on surface A with density σ_A^*. Assuming that $\mu_1 \gg \mu_2$, the pole density on surface A can be written as follows:

$$\sigma_A^* = \sigma_a^* + \sigma_b^* = \frac{1}{\mu_2} (B_{nb} - B_{na}) \qquad (5.2.19)$$

In order to represent σ_A^* in terms of the longitudinal field components, we apply Maxwell's equation, $\nabla \cdot \mathbf{B} = 0$, within sheet A, so that

$$\int_a^b \frac{\partial \mathbf{B}_n}{\partial n} dn = -\int_a^b \left(\frac{\partial \mathbf{B}_k}{\partial k} + \frac{\partial \mathbf{B}_l}{\partial l}\right) dn = -\int_a^b \nabla_A \cdot \mathbf{B} \, dn \qquad (5.2.20)$$

where $\nabla_A \cdot$ is the surface divergence operating on A. Assuming that the longitudinal field in A is independent of coordinate n, equation (5.2.20) acquires the following form:

$$B_{nb} - B_{na} = -t\,\nabla_A \cdot \mathbf{B} \qquad (5.2.21)$$

Substituting equation (5.2.21) into equation (5.2.19), applying the constitutive relation $\mathbf{B} = \mu_1 \mathbf{H}$ and using the notation $g^* = \mu_1 t$, we then obtain:

$$\sigma_A^* = -\frac{1}{\mu_2} \nabla_A \cdot g^* \mathbf{H} \qquad (5.2.22)$$

The pole density σ_C^* on the thin edges C of the sheet is as follows:

$$\sigma_C^* = \left(\frac{1}{\mu_2} - \frac{1}{\mu_1}\right) B_C \qquad (5.2.23)$$

where B_C is the component of the longitudinal magnetic flux density (B_k, B_l) perpendicular to edge C. As the sheet is very thin, the pole distributions on the edges can be considered as line poles with line density Q_C^*:

$$Q_C^* = \sigma_C^* t \qquad (5.2.24)$$

Substituting $B_C = \mu_1 H_C$, where H_C is the component of the longitudinal magnetic field (H_k, H_l) perpendicular to edge C, we obtain from equations (5.2.23) and (5.2.24):

$$Q_C^* = \frac{1}{\mu_2} g^* H_C \qquad (5.2.25)$$

Taking into consideration that $\sigma^* = \sigma_A^* + \sigma_C^*$ and applying equation (5.2.24), equation (5.2.17) can be written in the following form:

$$U^* = U_P^* + \int_A G\sigma_A^* \, dA_0 + \oint_C GQ_C^* \, dc_0 \qquad (5.2.26)$$

where c is the contour obtained by reducing the edges C to vanishingly thin lines. Substituting equations (5.2.22) and (5.2.25) into equation (5.2.26), we obtain for the potential:

$$U^* = U_P^* + \frac{1}{\mu_2}\left(\int_A -G\nabla_A \cdot g^* \mathbf{H} \, dA_0 + \oint_C Gg^* H_C \, dc_0\right) \qquad (5.2.27)$$

Applying the partial differentiation rule to the first integrand in equation (5.2.27) and Gauss's theorem to the second integral, and substituting

$$(H_k, H_l) = -\nabla_A U^* \qquad (5.2.28)$$

equation (5.2.27) becomes:

$$U^*(\mathbf{R}) = U_P^*(\mathbf{R}) + \frac{1}{\mu_2}\int_A -g^*(\mathbf{R}_0)\nabla_{A0} U^*(\mathbf{R}_0) \cdot \nabla_{A0} G(\mathbf{R}, \mathbf{R}_0)\,dA_0 \qquad (5.2.29)$$

where ∇_{A0} denotes the differentiation on surface A at the source point (k_0, l_0).

The primary potential U_P^* in equation (5.2.29) can be obtained by taking into consideration that for geophysical applications, the primary field is generally a constant vector, $\mathbf{H}_P$. Representing $\mathbf{H}_P$ by its components in Cartesian coordinates x, y, z:

$$\mathbf{H}_P = \mathbf{H}_{Px} + \mathbf{H}_{Py} + \mathbf{H}_{Pz}$$

and using the relation $-\nabla U_P^* = \mathbf{H}_P$, the primary potential can be written in the following form:

$$U_P^*(x, y, z) = -H_{Px}x - H_{Py}y - H_{Pz}z \qquad (5.2.30)$$

The solution of the magnetic thin sheet problem now proceeds as follows. The numerical solution for equation (5.2.29) can be obtained by a procedure similar to that used in Section 3.3.6 for solving the analogous equation, equation (3.3.60), for the electric potential of a thin conductor. The potential values U^* on surface A, obtained as the solution of equation (5.2.29), are then substituted into the left-hand side of equation (5.2.17). The equation thus obtained permits the pole density σ^* on surface A to be found. Finally, the magnetic anomaly $\mathbf{H}_S$, generated by the obtained pole density σ^*, can be calculated at an arbitrary point $\mathbf{R}$ of the model by using the equation:

$$\mathbf{H}_S(\mathbf{R}) = \int_A -\nabla G(\mathbf{R}, \mathbf{R}_0)\sigma^*(\mathbf{R}_0)dA_0 \qquad (5.2.31)$$

5.3 DEMAGNETIZATION AND LOW-SUSCEPTIBILITY MODELS

5.3.1 Introduction

When a ferromagnetic body is brought into a magnetizing (primary) field, a distribution of magnetic poles is generated on the surface of the body (see Fig. 5.2). The density of the poles depends upon the susceptibility of the body and the component of the magnetic field normal to the surface. The secondary field, generated by this pole distribution, reduces the field strength in the body, thus causing an effect known as **demagnetization**.

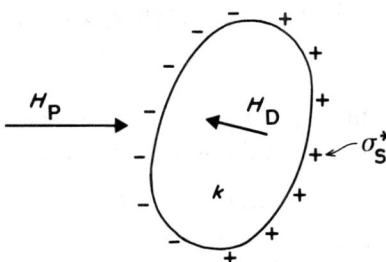

Figure 5.2 Demagnetizing field $\mathbf{H}_D$ generated in a magnetized body by surface pole distribution. $\mathbf{H}_P$ is the primary field.

When a magnetic field problem is solved by using an integral equation technique, the demagnetization is automatically taken into account. In practical geophysical interpretation, however, solving an integral equation by the standard numerical methods described in Section 1.3 is often too tedious to be part of automated or interactive curve fitting. Therefore, in modelling geological structures with low susceptibility values, simplified techniques are generally used when solving the integral equations.

It has been demonstrated by Eskola and Tervo [41, 42] that, if the susceptibility k of an anomalous body situated in a non-magnetized environment is lower than 0.01 SI, the surface pole density on the body can be obtained by simply substituting the total magnetic field in the body by the primary field. Using this approximation, the surface pole density on the body can be directly obtained from equation (4.4.7) by substituting $B_n = \mu H_{Pn}$. When the susceptibility in the body is higher than 0.01 SI and when the magnetic surveys aim at delineating the shape of the body, the secondary sources for susceptibility values $1 > k > 0.01$ SI can be determined by using the demagnetization factor. For susceptibility values higher than 1 SI, for bodies having other than a second-degree surface, the source density is strongly influenced by certain special geometrical characteristics of the body, such as its edges. These edge effects result in an increasingly non-homogeneous secondary field in the body. Therefore, for susceptibility values higher than 1 SI, the demagnetization factor method, which only describes the mean secondary field in a body, cannot be used for determining the secondary sources.

In contrast to the modelling problems associated with geological interpretation, the dimensions of the samples used in petrophysical laboratory measurements are often small compared with the measurement distance, and so detailed variations in the magnetization of the sample have negligible effect on the secondary field observed. It is sufficient that the magnetic moment measured be uniquely related to the magnetic properties of the sample material. In such cases the demagnetization factor, if appropriately determined, can be used even when the susceptibility is higher than 1 SI. In the following, we shall define the concept of the demagnetization factor and give an equation by which it can be calculated numerically. Further, we shall briefly consider simplified techniques for solving integral equations when interpreting geological formations whose susceptibilities are lower than 1 SI.

5.3.2 The demagnetization factor

Let us consider a magnetized body V with constant susceptibility k, situated in a homogeneous primary field $\mathbf{H}_P$. We define the demagnetization tensor $\mathbf{N}$ of body V as follows:

$$\mathbf{m} = \mathbf{m}_P(\mathbf{I} + \mathbf{N}k)^{-1} \qquad (5.3.1)$$

where $\mathbf{I}$ is the identity tensor, $\mathbf{m}$ is the magnetic moment of the body and $\mathbf{m}_P$

112   Magnetic methods

is the primary magnetic moment, that is, the moment that the body would have if the field strength in it were equal to the primary field strength $\mathbf{H}_P$.

Suppose that the demagnetization tensor $\mathbf{N}$ is symmetric, and that its principal axes are a, b and c. Then equation (5.3.1) can be expressed in the system of coordinates a, b, c in component form as follows:

$$m_i = \frac{m_{Pi}}{1 + N_i k} \tag{5.3.2}$$

where N_i is the demagnetization tensor in the direction $i = a, b, c$. It is known as the **demagnetization factor** of body V in this direction.

The i-component of the magnetic moment can also be obtained from the following equation:

$$m_i = \int_S x_i \sigma_S^* dS \tag{5.3.3a}$$

where σ_S^* is the magnetic surface pole density on surface S of body V due to the magnetization in V, and x_i is the coordinate along the i-axis. Correspondingly, the primary magnetic moment can be written in the following form:

$$m_{Pi} = \int_S x_i \sigma_P^* dS \tag{5.3.3b}$$

where the surface pole density corresponding to the primary field is σ_P^*, hence

$$m_{Pi} = -\int_S x_i k H_{Pn} dS = k H_{Pi} V \tag{5.3.4}$$

H_{Pi} is the i-component of the primary field and V denotes the volume of the body. From equations (5.3.2), (5.3.3a) and (5.3.4) it follows that

$$N_i = \frac{H_{Pi} V}{\int_S x_i \sigma_S^* dS} - \frac{1}{k} \tag{5.3.5}$$

The unknown surface pole density in equation (5.3.5) can be obtained from equation (5.2.3), after which the demagnetization factor N_i can be calculated using equation (5.3.5).

We see immediately from equation (5.3.5) that the demagnetization factor N_i depends on susceptibility k, but the dependence is not a simple one because σ_S^* is also a function of k.

Fig. 5.3 illustrates the demagnetization factors for a rectangular prism and a cube, when the susceptibility varies in the range 0.1–900 SI. The final values of the factors were obtained by solving equation (5.3.5) for a series of different numbers of boundary elements. The values thus obtained were then extrapolated to an infinite number of elements by fitting a proper extrapolation function to these values [41, 42].

The demagnetization factors shown in Fig. 5.3 decrease monotonically with

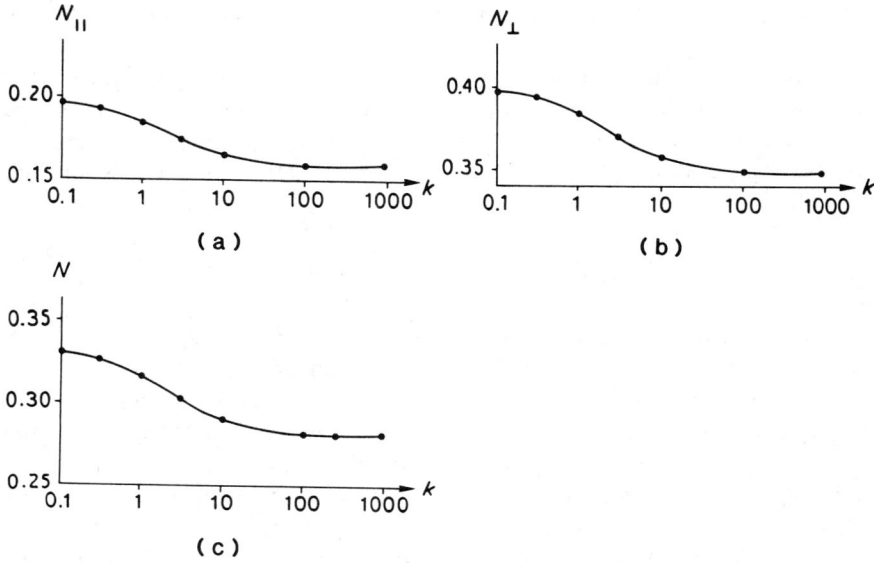

Figure 5.3 Demagnetization factor as a function of susceptibility, for a rectangular prism (dimensional ratios 1:1:2) (a) parallel to the long edge, (b) parallel to a short edge, and (c) for a cube. After [41, 42].

increasing susceptibility. The curves show considerable dispersion in the region $0.1 < k < 10$ SI and approach definite values, as k approaches zero and infinity.

The numerical results obtained were verified by measuring the demagnetization factor values for ferrite ($k = 900$ SI) samples having the same shape as those used in the numerical calculations. The results obtained by the measurements and those obtained from equation (5.3.5) by extrapolation agreed to within 1%. The efficiency and necessity of the extrapolation are apparent from the fact that the numerical demagnetization factor of the cube at $k = 900$ SI, obtained directly from equation (5.3.5) and using 1350 square boundary elements of equal size, was as much as 10% higher than the corresponding measured value.

5.3.3 Low-susceptibility models

Forward modelling by complete numerical solution of an integral equation is generally too tedious to be applied as a basis for automatic or interactive optimizing procedures of quantitative interpretation. To make the model calculations reasonably easy, it is usually assumed that the anomalous body is homogeneously magnetized and is situated in a non-magnetic medium. In fact, the magnetization is homogeneous only in bodies bounded by second-degree surfaces. For practical purposes, however, the homogeneity assumption can be considered to be sufficiently accurate whenever the susceptibility does not exceed 1 SI.

When deriving formulas for calculating anomalies caused by simple-shaped bodies, volume dipole distributions (see Section 4.5) rather than surface pole distributions are commonly used as the starting system, as long as homogeneity of magnetization is assumed. Using the relation $\mathbf{H} = -\nabla U^*$, we obtain from equation (4.5.7):

$$\mathbf{H}(\mathbf{R}) = -\nabla \int_V \mathbf{M} \cdot \nabla_0 G(\mathbf{R}, \mathbf{R}_0) \mathrm{d}V_0 \tag{5.3.6}$$

where $\mathbf{M}$ is the constant magnetization in the anomalous body V. The integral in equation (5.3.6) is completely equivalent to equation (A1.9) in Parasnis [9].

In determining $\mathbf{M}$, use is made of the relation $\mathbf{M} = k\mathbf{H}$, where k is the constant susceptibility in body V. For k values lower than 0.01 SI, the magnetization can be obtained simply by substituting $\mathbf{M} = k\mathbf{H}_P$, where $\mathbf{H}_P$ is the primary field. For susceptibility values $0.01 < k < 1$ SI, $\mathbf{M}$ can be determined by using the demagnetization factors as follows.

Let the demagnetization factors along the principal directions a, b and c be N_a, N_b and N_c.

Representing $\mathbf{M}$ in terms of its components

$$\mathbf{M} = \mathbf{M}_a + \mathbf{M}_b + \mathbf{M}_c \tag{5.3.7}$$

we can write $\mathbf{M}$ in terms of the corresponding components of the primary field $\mathbf{H}_P$ as follows:

$$M_i = \frac{kH_{Pi}}{1 + N_i k}, \quad i = a, b, c \tag{5.3.8}$$

Substituting equations (5.3.7) and (5.3.8), the anomaly generated by body V can be calculated using equation (5.3.6).

If one prefers to calculate the anomaly by applying the surface integral representation, use is made of equation (5.2.2). Taking into consideration that $\mathbf{M}_2 = 0$, we conclude from equations (4.4.5) and (4.5.4) that $\sigma_S^* = M_n$ where M_n is the normal component of magnetization $\mathbf{M}$ on the surface S of body V, hence

$$\mathbf{H}(\mathbf{R}) = -\nabla \int_S G(\mathbf{R}, \mathbf{R}_0) M_n \mathrm{d}S_0 \tag{5.3.9}$$

The magnetization $\mathbf{M}$ can be obtained by the procedures defined by equations (5.3.7) and (5.3.8).

Finally let us consider how the simplified formula for calculating the magnetic anomaly of a thin sheet (see [9, Appendix 5]) is deduced from the more exact formula given in Section 5.2.1. The scalar potential of the anomalous field is obtained from equation (5.2.17):

Demagnetization and low-susceptibility models 115

$$U^*(\mathbf{R}) = \int_A G(\mathbf{R}, \mathbf{R}_0) \sigma^*(\mathbf{R}_0) dA_0 \quad (5.3.10)$$

where the surface pole density σ^* consists of both the net pole density σ_A^* distributed on the large faces of the sheet, and the density σ_C^* distributed on the thin edges. Assuming that the magnetization **M** is homogeneous in the sheet, and substituting $\mu_2 = \mu_0$, the surface pole densities σ_a^* and σ_b^* given in equation (5.2.18), and distributed along the opposite faces of the sheet, are of equal magnitude but of opposite sign, and hence

$$\sigma_a^* = -\sigma_b^* = \left(\frac{1}{\mu} - \frac{1}{\mu_0}\right) B_\perp \quad (5.3.11)$$

where $B_\perp$ is the component of **B** normal to the faces of sheet A. According to this equation, the net pole density σ_A^* on the faces is zero, and the only contribution of pole distributions σ_a^* and σ_b^* is to form a homogeneous double layer distributed on A. Fig. 3.7(b), which represents the analogous case of a thin conductor, shows that for a high-susceptibility model, this condition is not in fact fulfilled. Check computations for a thin conductor with high susceptibility show that, in reality, the most significant sources of the secondary potential are the net surface charges distributed on A.

The density τ of the double layer formed by the pole distributions described by equation (5.3.11) can be written as follows

$$\tau = t\left(\frac{1}{\mu_0} - \frac{1}{\mu}\right) B_\perp \quad (5.3.12)$$

where t is the thickness of the sheet. Substituting $B_\perp = (\mu\mu_0/(\mu-\mu_0)) M_\perp$ into equation (5.3.12), we obtain for the dipole moment density:

$$\tau = tM_\perp \quad (5.3.13)$$

The density of poles distributed on the thin edges C of the sheet can be obtained from equation (5.2.23):

$$\sigma_C^* = \left(\frac{1}{\mu_0} - \frac{1}{\mu}\right) B_C \quad (5.3.14)$$

where B_C is the component of the longitudinal magnetic flux $B_\parallel$ normal to edge C. Substituting $B_C = (\mu\mu_0/(\mu-\mu_0)) M_C$ and considering the thickness t of the sheet to be very small, the pole distributions on the edges can be treated as line poles with line density Q_C^*:

$$Q_C^* = tM_C \quad (5.3.15)$$

The potential U^* of the anomalous field can be calculated from the equation

$$U^* = \int_A \tau \frac{\partial G_0}{\partial n_0} dA_0 + \int_C Q_C^* G_0 dc_0 \quad (5.3.16)$$

where $G_0 = 1/4\pi |\mathbf{R} - \mathbf{R}_0|$ and n is the coordinate of the axis perpendicular to

116 *Magnetic methods*

surface A. Substituting equations (5.3.13) and (5.3.15) and using the relation $\mathbf{H} = -\nabla U^*$ equation (5.3.16) becomes

$$\mathbf{H} = \nabla \int_A -M_\perp t \frac{\partial G_0}{\partial n_0} \, dA_0 - \nabla \int_C M_C t G_0 \, dc_0 \qquad (5.3.17)$$

Applying Gauss's law to surface A:

$$\int_C M_C G_0 \, dc_0 = \int_A \mathbf{M}_\| \cdot \nabla_{A_0} G_0 \, dA_0$$

it is evident that equation (5.3.17) is equivalent to equation (A5.3) in Parasnis [9]. Approximating the transverse and the longitudinal demagnetization factors in the sheet by $N_\perp = 1$ and $N_\| = 0$, we obtain the transverse and longitudinal magnetizations as $M_\perp = kH_{\mathrm{P}\perp}/(1 + k)$ and $M_C = kH_{\mathrm{PC}}$.

5.4 NUMERICAL APPLICATIONS

In the preceding three sections we concentrated on theoretical aspects of magnetic modelling using the integral equation technique. In the following,

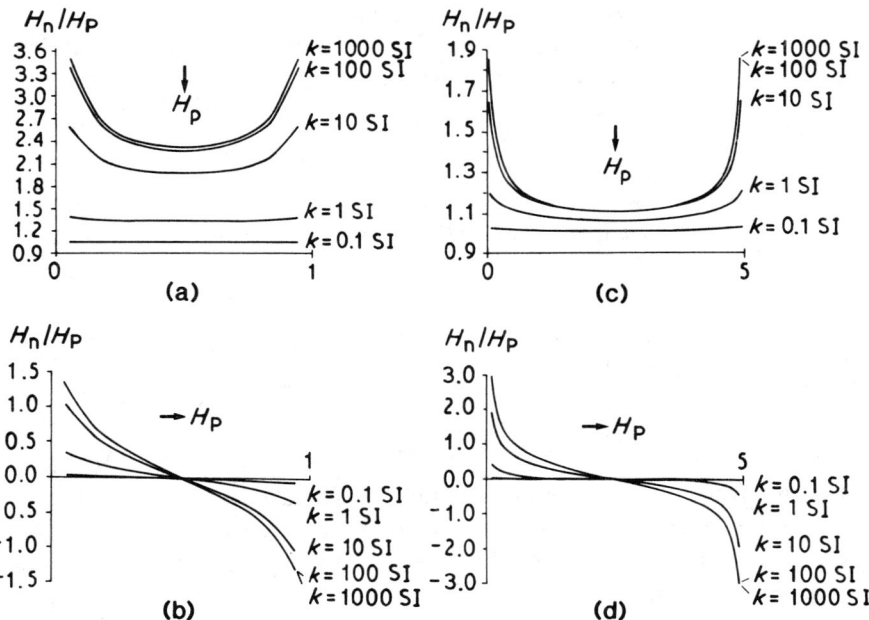

Figure 5.4 Component of magnetic field strength normal to the external surface of a $2\frac{1}{2}$-dimensional rectangular prism with susceptibility as a parameter. Dimensional ratio of cross-section is 1:5. The profiles (a) and (b) run along the shorter side, and (c) and (d) along the longer side of the cross-section. The primary field in (a) and (c) is normal to and in (b) and (d) tangential to the profile. After [43].

Numerical applications 117

we shall apply this technique to calculate a series of numerical results that characterize the magnetic poles distributed on the faces of a magnetized $2\frac{1}{2}$-dimensional prism, and the magnetic anomalies generated by these pole distributions. The results should facilitate understanding of the physical phenomena involved in the theoretical formulations.

Figure 5.4 illustrates the external normal component of the magnetic field, H_n, along the sides of the cross-section of the faces of a $2\frac{1}{2}$-dimensional rectangular prism, when the susceptibility of the prism varies. The dimensional ratio of the prism's cross-section is 1:5, and the strike length is 100 times the longer dimension of the cross-section. The field problem is solved by using equations (5.2.10)–(5.2.14), and for numerical solution the cross-sectional contour of the prism is divided into 108 subsections of equal length. The relation between H_n and the surface pole density σ_s^* can be expressed by the formula $\sigma_s^* = kH_n/(1+k)$.

When susceptibility k is less than 1 SI, the effect of the singularities at the corners is significant only in their immediate vicinity, and the field along each side of the cross-section rectangle can be considered as homogeneous. Hence, the concept of the demagnetization factor can be used in calculating the anomaly caused by the body. However, when k is greater than 1 SI, the contribution of the singularities is so strong that the anomaly caused by the body cannot be modelled using the demagnetization factor technique.

The linear singularities at the prism edges, responsible for making the magnetization in the prism non-homogeneous, arise due to magnetostatic coupling between the prism faces intersecting along these edges. When the primary field is directed so that the interaction between the surface poles distributed on contiguous faces is constructive, as it is in the case under consideration, the pole densities tend to infinity towards the edges. Since the primary field is tangential to the profiles in the cases depicted by Figs 5.4(b) and 5.4(d), the normal components of the field are entirely due to interaction between the pole distributions on the prism faces.

Figure 5.5 represents the vertical-field magnetic anomalies of the prism defined in Fig. 5.4. The anomalies are calculated by using equations (5.2.10)–(5.2.14), and the demagnetization factor is obtained from equation (5.3.5). The primary field is vertical and the susceptibilities are 0.1 SI or 10 SI. For the low susceptibility value 0.1 SI (Figs 5.5(a) and 5.5(c)), the anomalies obtained by the demagnetization technique and by solving the integral equation with different degrees of approximation are practically the same, indicating that the application of the demagnetization factor technique is reasonable in such cases. On the other hand, for the susceptibility value 10 SI (Figs 5.5(b) and 5.5(d)), the two anomalies differ considerably from each other. The anomalies calculated using demagnetization factors are inaccurate in both shape and intensity. This is mainly due to the strong effect of the linear singularities at the prism edges. These cannot be simulated with the aid of the demagnetization factor.

118 *Magnetic methods*

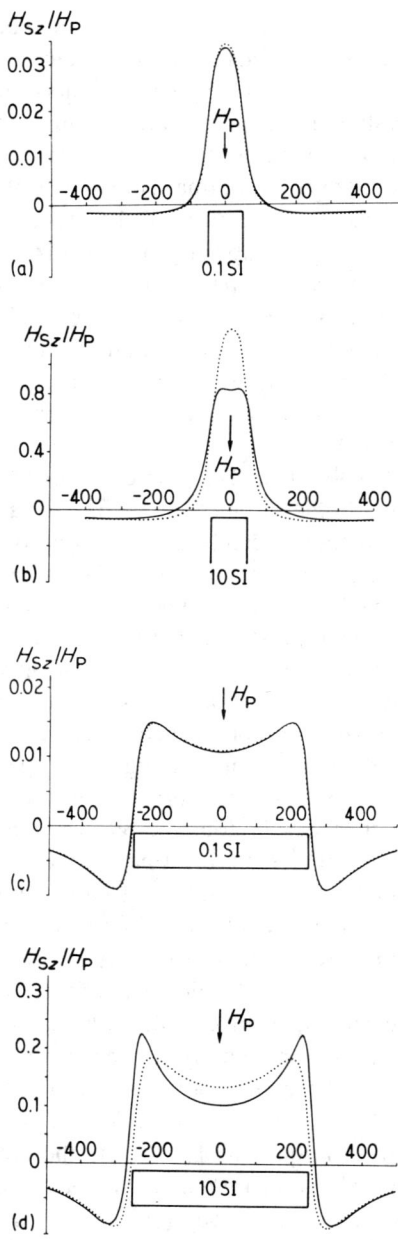

Figure 5.5 Vertical anomaly of the $2\frac{1}{2}$-dimensional prism of Fig. 5.4, magnetized by vertical primary field. Anomalies obtained from integral equation by using the method of subsections (solid line), and the demagnetization factor technique (dashed line). Prism in vertical position, (a) $k = 0.1$ SI, (b) $k = 10$ SI; and in horizontal position, (c) $k = 0.1$ SI, (d) $k = 10$ SI.

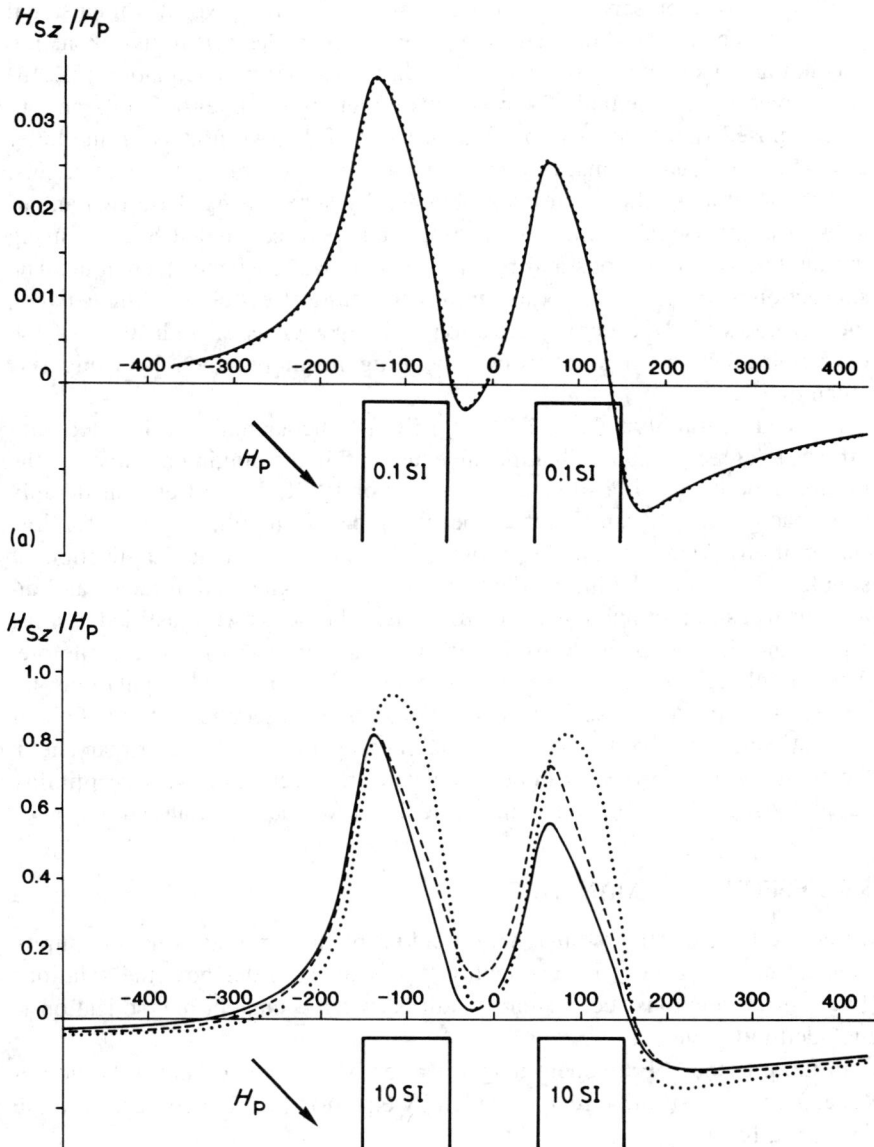

Figure 5.6 Vertical anomaly for two adjacent vertical $2\frac{1}{2}$-dimensional prisms as in Fig. 5.4, with interaction between the prisms taken into account (solid line), and interaction neglected. For the latter case, partial anomalies due to the individual prisms have been calculated using the method of subsections (dashed line) and the demagnetization factor (dotted line). Inclination of the primary field 45°. (a) $k = 0.1$ SI, and (b) $k = 10$ SI.

Figure 5.6 represents vertical-field anomalies for two $2\frac{1}{2}$-dimensional prisms, when the magnetostatic coupling between the prisms is taken into account and also when it is disregarded. In the former case, equation (5.2.13) was solved by the method of subsections by taking the integration domain to be composed simultaneously of both surfaces of the two prisms. In the latter case, the individual anomalies were first calculated separately for both prisms, and the total anomalies were then obtained by superposing these two anomalies. The individual anomalies of the prisms were calculated both by using the method of subsections and by the demagnetization factor technique. The subsection result allows us to take into account the effect of coupling between the prisms, while the demagnetization technique gives an indication of the overall inaccuracy in a magnetic modelling result obtained by using this technique.

For a susceptibility value 0.1 SI (Fig. 5.6(a)), the anomalies calculated with different degrees of approximation are equal within the plotting accuracy of the figure. In contrast, for a susceptibility value of 10 SI, they differ considerably from each other. It can therefore be concluded from Fig. 5.6 that for low susceptibility values, including those of common rocks, the application of simple magnetic modelling methods using the demagnetization factor and direct superposition of anomalies due to separate bodies is well justified. On the other hand, if the susceptibility is very high, as for a massive magnetite ore, these simple approximation methods are too coarse for model calculations. Of course, if the bodies are so far from each other that the secondary field of either body at the location of the other is weak in comparison with the primary field, the anomaly of the system can be calculated, irrespective of the susceptibility value, by superposing the anomalies calculated for each separate body.

5.5 EFFECT OF REMANENCE

So far we have considered magnetic field problems for models having purely induced magnetization. In this section, we shall consider how the solutions given in the previous sections are modified when remanent magnetization is included in the model.

In the presence of remanent magnetization $\mathbf{M}_R$, assumed here to be homogeneous, the constitutive relation given by equation (4.2.1) generalizes to the following form:

$$\mathbf{H} = \frac{1}{\mu}\mathbf{B} - \frac{\mu_0}{\mu}\mathbf{M}_R \qquad (5.5.1)$$

From Gauss's law it then follows that the volume pole density σ^*, given in equation (4.3.3), is of the following more complicated form

$$\sigma^* = \mathbf{B} \cdot \nabla\left(\frac{1}{\mu}\right) - \mu_0 \mathbf{M}_R \cdot \nabla\left(\frac{1}{\mu}\right) \qquad (5.5.2)$$

Further, the relation between the surface pole density and the field vectors, given by equation (4.4.7), becomes

$$\sigma_S^* = \left(\frac{1}{\mu_2} - \frac{1}{\mu_1}\right) B_n - \mu_0 \left(\frac{M_{Rn2}}{\mu_2} - \frac{M_{Rn1}}{\mu_1}\right) \qquad (5.5.3)$$

When the magnetic field in a model having remanent magnetization is represented by means of the surface integral (see Fig. 4.1)

$$\mathbf{H}(\mathbf{R}) = \mathbf{H}_P(\mathbf{R}) - \int_S \nabla G(\mathbf{R}, \mathbf{R}_0) \sigma_S^*(\mathbf{R}_0) dS_0 \qquad (5.5.4)$$

the surface pole density σ_S^* to be substituted is that given by equation (5.5.3). The integral equation for solving σ_S^*, obtained from equation (5.5.4) by moving the calculation point $\mathbf{R}$ to surface S, is, instead of equation (5.2.3), of the following form:

$$\frac{\mu_1 + \mu_2}{2(\mu_1 - \mu_2)} \sigma_S^*(\mathbf{R}) + \frac{\mu_0}{(\mu_1 - \mu_2)} (M_{Rn1}(\mathbf{R}) - M_{Rn2}(\mathbf{R})) =$$

$$H_{Pn}(\mathbf{R}) - \int_S \frac{\partial G(\mathbf{R}, \mathbf{R}_0)}{\partial n} \sigma_S^*(\mathbf{R}_0) dS_0 \qquad (5.5.5)$$

When the contribution of remanence is included, the continuity condition on B_n across boundary A, which in equation (4.2.10b) corresponds to purely induced magnetization, becomes more general and the boundary conditions for determining Green's function G become the following:

$$G_3 = G_2$$
$$\mu_3 \frac{\partial G_3}{\partial n} - \mu_0 M_{Rn3} = \mu_2 \frac{\partial G_2}{\partial n} - \mu_0 M_{Rn2} \qquad (5.5.6)$$

on surface A.

Finally, let us consider the case in which a magnetized body V_1, bounded by surface S, is situated in a non-magnetized space V_2. Let the susceptibility in region V_1 be k and the remanent magnetization $\mathbf{M}_R$. The constitutive relations between the field vectors are thus:

$$\mathbf{B}_1 = \mu_0((1+k)\mathbf{H}_1 + \mathbf{M}_R), \quad \text{in } V_1$$

$$\mathbf{B}_2 = \mu_0 \mathbf{H}_2, \qquad \qquad \text{in } V_2$$

Applying Gauss's law according to equation (5.5.3) we obtain, for the surface pole density on S:

$$\sigma_S^* = kH_n + M_{Rn} \qquad (5.5.7)$$

Representing the magnetic field in terms of its primary and secondary parts,

122 Magnetic methods

$\mathbf{H} = \mathbf{H}_P + \mathbf{H}_S$, we can write for the secondary field:

$$\mathbf{H}_S(\mathbf{R}) = \int_S \nabla G_0(\mathbf{R}, \mathbf{R}_0) \, (kH_{Pn}(\mathbf{R}_0) + M_{Rn}(\mathbf{R}_0)) \, dS_0 +$$

$$\int_S \nabla G_0(\mathbf{R}, \mathbf{R}_0) \, kH_{Sn}(\mathbf{R}_0) \, dS_0 \qquad (5.5.8)$$

The first integral in equation (5.5.8) can be considered as the 'primary' part of the anomalous field, generated by the pole distributions due to the primary field and the remanent magnetization. The second integral, which is a consequence of the first integral, represents the 'secondary' part of $\mathbf{H}_S$ and is responsible for the demagnetization and other, higher-order interaction effects in the pole distributions on surface S.

It can be seen from the first integral of equation (5.5.8) that, in causing the anomalous field of the simple model under consideration, the contribution of the remanent magnetization $\mathbf{M}_R$ is similar to that of the product, $k\mathbf{H}_P$ of the primary field and the susceptibility. Consequently, in the low-susceptibility models considered in Section 5.3.3, the contribution of remanence can be taken into account by substituting equation (5.3.8) by

$$M_i = \frac{kH_{Pi} + M_{Ri}}{1 + N_i k}, \quad i = a, b, c \qquad (5.5.9)$$

Here the M_i now represent the total magnetizations, $M_i = kH_i + M_{Ri}$, $i = a, b, c$.

6

Electromagnetic methods

6.1 INTRODUCTION

In the previous four chapters we considered electric and magnetic methods whose mathematical treatment was based on the theory of static fields. It was stressed that the physical sources for the electric field are charges and those for the magnetic field are atomic or macroscopic electric currents.

In this chapter we shall consider methods which are based on the electromagnetic induction phenomenon associated with time-varying electromagnetic fields in a medium. The sources of the electromagnetic field are charges at rest and in motion defined macroscopically in terms of the charge density σ and current density $\mathbf{j}$. In addition to being generated by charges and currents, the basic field vectors, the electric field $\mathbf{E}$ and the magnetic flux density $\mathbf{B}$ are also interrelated in space and time. This interrelation and the relation with the field sources are represented by Maxwell's equations (2.1.1) and (2.1.2).

The physical phenomena associated with time-varying electromagnetic fields are more complicated than those associated with the static electric and magnetic fields so that the formal representation of time-varying fields is of considerable complexity. At the procedural level, however, the methods by which Maxwell's equations are transformed into boundary-value problems and subsequently converted into integral equations are very similar for time-dependent and the static systems.

A number of different integral equation solutions for electromagnetic scattering problems have been published in the geophysical literature during the past two decades, dealing with general three-dimensional models as well as with more restricted special geometries. It is not possible to give thorough descriptions of all the relevant formulations in this book since our aim is merely to give an introduction to the methods used in generating integral equation solutions for electromagnetic problems. First, we shall consider how the time-varying electric and magnetic boundary-value problems are converted into integral equations by applying the vector-dyadic Green's second identity.

6.2 BOUNDARY-VALUE PROBLEMS FOR ELECTROMAGNETIC FIELDS

For electromagnetic problems, time dependence can generally be treated most conveniently in the frequency domain. The time-domain response can then be obtained from the corresponding frequency-domain response by an inverse Fourier transformation. Hence, we shall first reproduce Maxwell's equations in the frequency domain by assuming the time dependence $\exp(i\omega t)$ of the system, where $\omega = 2\pi f$ is the angular frequency and $i^2 = -1$. Accordingly, substituting in equations (2.1.1)–(2.1.5) the factor $i\omega$ for the operator $\partial/\partial t$, we obtain Maxwell's equations in the frequency domain as follows:

$$\nabla \times \mathbf{E} = -i\omega\mu\mathbf{H} \tag{6.2.1}$$

$$\nabla \times \mathbf{H} = (g + i\varepsilon\omega)\mathbf{E} + \mathbf{j}^s \tag{6.2.2}$$

$$\nabla \cdot \mathbf{B} = 0 \tag{6.2.3}$$

$$\nabla \cdot \mathbf{D} = \sigma_f \tag{6.2.4}$$

$$\nabla \cdot \mathbf{j} = -i\omega\sigma_f \tag{6.2.5}$$

where σ_f is the free volume charge density, g is conductivity of the medium and $\mathbf{j}^s$ stands for the density of source current flowing in the transmitter system.

Next we shall formulate the boundary-value problems for time-varying electric and magnetic fields on the basis of equations (6.2.1)–(6.2.5). Several sets of vector and/or scalar potential functions for electromagnetic field vectors have been defined and applied in the literature in order to facilitate the analytical solution of field problems. In this treatment however, we represent the boundary-value problems in terms of the field vectors $\mathbf{E}$ and $\mathbf{H}$ themselves. Appropriate potential functions may then be used when analytically solving the dyadic Green's functions for the models under consideration.

6.2.1 Electric field

To formulate the boundary-value problem for $\mathbf{E}$ from Maxwell's equations, we incorporate the physical properties of the medium into the system by the constitutive relations $\mathbf{j} = g\mathbf{E}$, $\mathbf{B} = \mu\mathbf{H}$ and $\mathbf{D} = \varepsilon\mathbf{E}$, where conductivity g, permeability μ and permittivity ε are assumed to be isotropic and the medium linear. By taking the curl and eliminating $\mathbf{H}$ in equations (6.2.1) and (6.2.2), we obtain the non-homogeneous vector wave equation for $\mathbf{E}$:

$$\nabla \times \nabla \times \mathbf{E} - k^2 \mathbf{E} = -i\omega\mu \mathbf{j}^s + \frac{\nabla \mu}{\mu} \times \nabla \times \mathbf{E} \tag{6.2.6}$$

where k is the propagation constant, defined as follows:

$$k^2 = -i\omega\mu(g + i\varepsilon\omega) \tag{6.2.7}$$

The behaviour of the electromagnetic field at physical boundaries can be deduced from Maxwell's equations, which imply that the tangential components of the electric and magnetic fields are continuous across the boundaries, hence:

$$\mathbf{n} \times \mathbf{E}_j = \mathbf{n} \times \mathbf{E}_{j+1} \tag{6.2.8}$$

$$\mathbf{n} \times \mathbf{H}_j = \mathbf{n} \times \mathbf{H}_{j+1} \tag{6.2.9}$$

where $\mathbf{n}$ denotes the normal unit vector of the boundary surface between media j and $j+1$. Equations (6.2.8) and (6.2.9) also guarantee that the normal components of current density $\mathbf{j}$ and magnetic flux density $\mathbf{B}$ are continuous across the boundary.

The behaviour of time-dependent electric and magnetic fields at infinity departs considerably from that of the static fields. This behaviour can be represented by the following radiation condition, taken, with a slight modification, from [44]:

$$\nabla \times \mathbf{E} - k\frac{\mathbf{R}}{R} \times \mathbf{E} = 0\left(\frac{1}{R}\right) \tag{6.2.10}$$

$$\nabla \times \mathbf{H} + k\frac{\mathbf{R}}{R} \times \mathbf{H} = 0\left(\frac{1}{R}\right) \tag{6.2.11}$$

as the calculation point $\mathbf{R}$ approaches infinity. It is sufficient to impose only one of the radiation conditions (6.2.10) and (6.2.11) in order completely to characterize the behaviour of the electromagnetic fields. The radiation condition implies that the time-varying electric and magnetic fields approach zero as $1/R$, as R approaches infinity. This is to be compared with the static fields which approach zero as $1/R^2$. The radiation condition ensures further that, at a great distance, the transport of electromagnetic energy is directed towards infinity. More detailed consideration of the radiation condition can be found, for example, in [45].

Let us consider the boundary-value problem defined by equations (6.2.6)–(6.2.11) for a model illustrated in Fig. 6.1, where V_1 is considered to be an anomalous region. To convert the boundary-value problem into an integral equation, we rewrite the partial differential equation (6.2.6) in a more appropriate form by partitioning the total electric field $\mathbf{E}$ into the primary and the secondary components $\mathbf{E}_P$ and $\mathbf{E}_S$:

$$\mathbf{E} = \mathbf{E}_P + \mathbf{E}_S \tag{6.2.12}$$

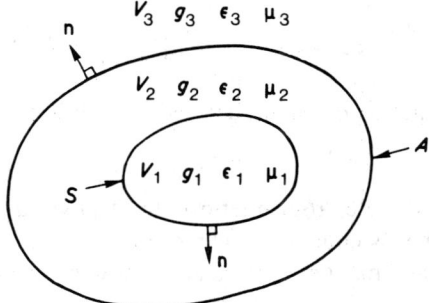

Figure 6.1 Three-dimensional structural model for electromagnetic problems.

The primary field $\mathbf{E}_P$ is defined as the field that would prevail in the model if the region V_1 were not anomalous. Assume that the physical properties g_2, ε_2 and μ_2 in region V_2, and g_3, ε_3 and μ_3 in region V_3 are constant, while g_1, ε_1 and μ_1 in the anomalous region V_1 may be variable. The primary field thus satisfies the wave equation:

$$\nabla \times \nabla \times \mathbf{E}_P - k_j^2 \mathbf{E}_P = -i\omega\mu_j \mathbf{j}^S \qquad (6.2.13)$$

where $k_j^2 = -i\omega\mu_j(g_j + i\omega\varepsilon_j)$. Subscript j takes values 2 or 3 such that $j = 2$ when the calculation point $\mathbf{R}$ is in either V_1 or V_2, and $j = 3$ when it is in V_3.

Writing $\Delta k^2 = k_1^2 - k_2^2$, and substituting equation (6.2.12) into equation (6.2.6), the wave equation for the secondary field $\mathbf{E}_S$ can be written in the following form:

$$\nabla \times \nabla \times \mathbf{E}_S - k_2^2 \mathbf{E}_S = \Delta k^2 \mathbf{E} + \frac{\nabla \mu_1}{\mu_1} \times \nabla \times \mathbf{E} \qquad \mathbf{R} \in V_1 \qquad (6.2.14)$$

$$\nabla \times \nabla \times \mathbf{E}_S - k_2^2 \mathbf{E}_S = 0 \qquad \mathbf{R} \in V_2 \qquad (6.2.15)$$

$$\nabla \times \nabla \times \mathbf{E}_S - k_3^2 \mathbf{E}_S = 0 \qquad \mathbf{R} \in V_3 \qquad (6.2.16)$$

Wave equations (6.2.14)–(6.2.16) and boundary conditions (6.2.8)–(6.2.11) define the boundary-value problem for the secondary field $\mathbf{E}_S$. We note that in equation (6.2.14), which represents the non-homogeneous wave equation for the field in anomalous region V_1, the source terms, represented by the right-hand side of the equation, are controlled by the unknown total field $\mathbf{E}$. Since the relation of the secondary field $\mathbf{E}_S$ to the total field $\mathbf{E}$ is uniquely known, the boundary-value problem defined is in a suitable form for conversion into an integral equation.

6.2.2 Magnetic field

As a consequence of the symmetry of Maxwell's equations with respect to the electric and magnetic fields $\mathbf{E}$ and $\mathbf{H}$, deduction of the boundary-value problem for the magnetic field is very similar to that for the electric field. Partitioning the total magnetic field **H** into the primary and secondary components, $\mathbf{H} = \mathbf{H}_P + \mathbf{H}_S$, taking the curl and eliminating **E** in equations (6.2.1) and (6.2.2) as in Section 6.2.1, we obtain the following wave equations for the secondary magnetic field $\mathbf{H}_S$ (see Fig. 6.1):

$$\nabla \times \nabla \times \mathbf{H}_S - k_2^2 \mathbf{H}_S = \Delta k^2 \mathbf{H} + \frac{\nabla(g_1 + i\omega\varepsilon_1)}{g_1 + i\omega\varepsilon_1} \times \nabla \times \mathbf{H}, \qquad \mathbf{R} \in V_1 \qquad (6.2.17)$$

$$\nabla \times \nabla \times \mathbf{H}_S - k_2^2 \mathbf{H}_S = 0, \qquad \mathbf{R} \in V_2 \qquad (6.2.18)$$

$$\nabla \times \nabla \times \mathbf{H}_S - k_3^2 \mathbf{H}_S = 0, \qquad \mathbf{R} \in V_3 \qquad (6.2.19)$$

Correspondingly, the wave equation for the primary field $\mathbf{H}_P$ can be obtained from equations (6.2.1) and (6.2.2) as follows:

$$\nabla \times \nabla \times \mathbf{H}_P - k_j^2 \mathbf{H}_P = \nabla \times \mathbf{j}^S \qquad (6.2.20)$$

When the source current $\mathbf{j}^S$ is flowing in a small loop, then it may be practical to substitute $\mathbf{j}^S$ by an equivalent distribution of magnetic dipoles with moment density $\mathbf{m}^S$. Writing Maxwell's equations (6.2.1) and (6.2.2) in the following form:

$$\nabla \times \mathbf{E} = -i\omega\mu(\mathbf{H} + \mathbf{m}^S) \qquad (6.2.21)$$

$$\nabla \times \mathbf{H} = (g + i\omega\varepsilon)\mathbf{E} \qquad (6.2.22)$$

and eliminating **E**, we obtain for the primary magnetic field $\mathbf{H}_P$:

$$\nabla \times \nabla \times \mathbf{H}_P - k^2 \mathbf{H}_P = k^2 \mathbf{m}^S \qquad (6.2.23)$$

Comparing equations (6.2.20) and (6.2.23), we obtain the relation between the density of the solenoidal current $\mathbf{j}^S$ and the magnetic moment density $\mathbf{m}^S$ equivalent to it:

$$\nabla \times \mathbf{j}^S = k^2 \mathbf{m}^S \qquad (6.2.24)$$

Wave equations (6.2.17)–(6.2.20), and boundary conditions (6.2.8)–(6.2.11) now define a boundary-value problem for the magnetic field, which is in a form appropriate for conversion into an integral equation.

6.3 GREEN'S DYADICS FOR ELECTROMAGNETIC BOUNDARY-VALUE PROBLEMS

In this section we shall describe how the electromagnetic boundary-value problem defined in Section 6.2 can be converted into an integral equation by applying the Green's function technique. The principle of conversion is similar to that applied in generating integral equation solutions for the static electric and magnetic fields.

Green's functions for time-varying systems are functionally more complicated than those for static systems, which is a consequence of the fact that the physical character of dynamic systems is more complicated than that of static systems. Furthermore, it can be seen from equations (6.2.14) and (6.2.17) that both the sources and the responses of time-varying systems are vectors, and hence, Green's functions of the time-varying systems are dyadic functions.

Conversion of electromagnetic boundary-value problems into integral equations is performed by applying the vector-dyadic second identity of Green, which is given below, following [46].

For the continuously differentiable vector function $\mathbf{F}$ and dyadic function $\tilde{\mathbf{D}}$, defined in region V, we have

$$\int_V [\mathbf{F} \cdot \nabla \times \nabla \times \tilde{\mathbf{D}} - (\nabla \times \nabla \times \mathbf{F}) \cdot \tilde{\mathbf{D}}] \, dV =$$

$$- \int_S [(\mathbf{n} \times \nabla \times \mathbf{F}) \cdot \tilde{\mathbf{D}} + (\mathbf{n} \times \mathbf{F}) \cdot \nabla \times \tilde{\mathbf{D}}] \, dS \qquad (6.3.1)$$

where S denotes the surface enclosing volume V and $\mathbf{n}$ the outward normal unit vector of S. In applying equation (6.3.1), we substitute for $\mathbf{F}$ the secondary electric and magnetic fields $\mathbf{E}_S$ and $\mathbf{H}_S$, and for $\tilde{\mathbf{D}}$ the electric and magnetic Green's dyadics $\tilde{\mathbf{G}}$ and $\tilde{\mathbf{G}}_m$.

As stated in connection with static field problems, it is reasonable to require that Green's function obeys the same basic relations as the fields generated by the sources. Calculation of electric or magnetic Green's dyadics is thus based on solving the boundary-value problem for the host structure in the model under consideration (that is, the structure when the anomalous region is not anomalous), with the source of the field being correspondingly an electric or magnetic unit dipole. As a consequence of the symmetry in Maxwell's equations with respect to the electric and magnetic fields, solutions of Green's dyadics for the electric and magnetic problems are completely analogous.

In the Cartesian system of coordinates, dyadic $\tilde{\mathbf{G}}$ can be expressed in component form as follows:

$$\tilde{\mathbf{G}} = g_{xx} \mathbf{xx} + g_{xy} \mathbf{yx} + g_{xz} \mathbf{zx}$$

$$+ g_{yx} \mathbf{xy} + g_{yy} \mathbf{yy} + g_{yz} \mathbf{zy}$$

$$+ g_{zx} \mathbf{xz} + g_{zy} \mathbf{yz} + g_{zz} \mathbf{zz} \qquad (6.3.2)$$

where $\mathbf{x}$, $\mathbf{y}$ and $\mathbf{z}$ denote the unit vectors in the directions of coordinate axes x, y and z, respectively. The components are defined so that, for example, the second term in the first row, $g_{xy} \mathbf{yx}$, represents the y component of the electric field at point $\mathbf{R}$, generated by the unit electric dipole located at point $\mathbf{R}_0$ and aligned parallel to the x-axis.

Green's dyadics

Referring to equations (6.2.14)–(6.2.16), the electric Green's dyadic $\tilde{\mathbf{G}}$ is required to satisfy the following non-homogeneous wave equation:

$$\nabla \times \nabla \times \tilde{\mathbf{G}}(\mathbf{R}, \mathbf{R}_0) - k^2 \tilde{\mathbf{G}}(\mathbf{R}, \mathbf{R}_0) = \tilde{\mathbf{I}} \delta(\mathbf{R} - \mathbf{R}_0) \qquad (6.3.3)$$

where $\mathbf{R}$ is the calculation point and $\mathbf{R}_0$ the source point. $\tilde{\mathbf{I}} = \mathbf{xx} + \mathbf{yy} + \mathbf{zz}$ is the idemfactor, and δ is Dirac's delta function. Propagation constant k takes the physical properties of region V_2 (see Fig. 6.1) if $\mathbf{R}$ is in either V_1 or V_2, and the properties of V_3, if $\mathbf{R}$ is in V_3.

The boundary conditions satisfied by the electric Green's dyadic $\tilde{\mathbf{G}}$ when the point $\mathbf{R}$ passes through the boundary A between media 2 and 3 of Fig. 6.1, are obtained from equations (6.2.8) and (6.2.9) [46]:

$$\mathbf{n} \times \tilde{\mathbf{G}}^{2,j} = \mathbf{n} \times \tilde{\mathbf{G}}^{3,j} \qquad (6.3.4)$$

$$\frac{1}{\mu_2} \mathbf{n} \times \nabla \times \tilde{\mathbf{G}}^{2,j} = \frac{1}{\mu_3} \mathbf{n} \times \nabla \times \tilde{\mathbf{G}}^{3,j} \qquad (6.3.5)$$

on surface A. The first superscript (2 or 3) in Green's dyadic refers to the location region of the calculation point $\mathbf{R}$, and the second ($j = 2, 3$) to that of the source point $\mathbf{R}_0$. The radiation condition satisfied by $\tilde{\mathbf{G}}$ can be obtained according to equation (6.2.10) in the following form:

$$\nabla \times \tilde{\mathbf{G}} - k \frac{\mathbf{R}}{R} \times \tilde{\mathbf{G}} = 0 \left(\frac{1}{R} \right) \qquad (6.3.6)$$

as $\mathbf{R}$ approaches infinity.

The magnetic Green's dyadic $\tilde{\mathbf{G}}_m$ of a time-varying magnetic field problem satisfies a wave equation similar to Green's dyadic for the electric field problem:

$$\nabla \times \nabla \times \tilde{\mathbf{G}}_m(\mathbf{R}, \mathbf{R}_0) - k^2 \tilde{\mathbf{G}}_m(\mathbf{R}, \mathbf{R}_0) = \tilde{\mathbf{I}} \delta(\mathbf{R} - \mathbf{R}_0) \qquad (6.3.7)$$

where k takes physical properties similar to those it has in equation (6.3.3). According to Fig. 6.1, the solution of equation (6.3.7) has to satisfy the following boundary conditions:

$$\mathbf{n} \times \tilde{\mathbf{G}}_m^{2,j} = \mathbf{n} \times \tilde{\mathbf{G}}_m^{3,j} \qquad (6.3.8)$$

$$\frac{1}{g_2 + i\omega\varepsilon_2} \mathbf{n} \times \nabla \times \tilde{\mathbf{G}}_m^{2,j} = \frac{1}{g_3 + i\omega\varepsilon_3} \mathbf{n} \times \nabla \times \tilde{\mathbf{G}}_m^{3,j} \qquad (6.3.9)$$

when the point $\mathbf{R}$ passes through surface A, and the radiation condition following from equation (6.2.11):

$$\nabla \times \tilde{\mathbf{G}}_m + k \frac{\mathbf{R}}{R} \times \tilde{\mathbf{G}}_m = 0 \left(\frac{1}{R} \right) \qquad (6.3.10)$$

as $\mathbf{R}$ approaches infinity.

130   *Electromagnetic methods*

The mixed Green's dyadics, that is, the dyadics transmitting the magnetic field due to unit electric sources and the electric field due to unit magnetic sources, can be obtained from $\tilde{\mathbf{G}}$ and $\tilde{\mathbf{G}}_m$ using equations (6.2.1) and (6.2.2):

$$\tilde{\mathbf{G}}_{me} = (i/\omega\mu)\nabla \times \tilde{\mathbf{G}}$$

$$\tilde{\mathbf{G}}_{em} = (1/g + i\omega\varepsilon)\nabla \times \tilde{\mathbf{G}}_m$$

The general properties of Green's dyadics $\tilde{\mathbf{G}}$ and $\tilde{\mathbf{G}}_m$ are very similar, the only differences being in the physical constants associated with the application of boundary conditions on surface A. Consequently, their general properties can be described by restricting consideration to only one of these functions, for example $\tilde{\mathbf{G}}$.

Green's dyadic $\tilde{\mathbf{G}}$ can be written as the following sum:

$$\tilde{\mathbf{G}} = \tilde{\mathbf{G}}_0 + \tilde{\mathbf{G}}_S \tag{6.3.11}$$

where $\tilde{\mathbf{G}}_0$ represents the singular particular solution of equation (6.3.3) which is known as the **whole-space Green's dyadic**. It satisfies the radiation condition (6.3.6) at infinity and is, together with its derivative, automatically continuous on surface A. The whole-space Green's dyadic $\tilde{\mathbf{G}}_0$ can be expressed in the following form [47; 48]:

$$\tilde{\mathbf{G}}_0 = \left(1 + \frac{1}{k^2}\nabla\nabla\cdot\right)G_0\tilde{\mathbf{I}} = \left(\tilde{\mathbf{I}} + \frac{1}{k^2}\nabla\nabla\right)G_0 \tag{6.3.12}$$

where the first term may be interpreted as the solenoidal part of the electric field generated by currents, and the second term as the laminar part generated by charges. The functional form of G_0 is as follows:

$$G_0(\mathbf{R}, \mathbf{R}_0) = \frac{e^{-ik|\mathbf{R}-\mathbf{R}_0|}}{4\pi|\mathbf{R}-\mathbf{R}_0|} \tag{6.3.13}$$

The second term, G_S, of equation (6.3.11) represents the non-singular secondary field generated by the scattering on boundary A. Since G_S carries the structural information of the space occupied by the anomalous body, its solution is the main problem in determining the Green's dyadic of a particular model.

The boundary-value problem for G_S can be solved analytically for only a very restricted group of structural models having boundaries that can be represented by constant coordinate values of a separable system of coordinates. In the case of geologically important models, easily computable Green's functions exist only for layered earth models, and hence, published numerical electromagnetic integral equation solutions are at present restricted to models of this kind.

Electric Green's dyadics for a general layered earth have been derived, for example, in [49] using Hertz potentials, and in [50] by using magnetic and electric Schelkunoff vector potentials. Electric Green's dyadics for a half-

Volume integral equations for 3-D fields 131

space in analytic form are given in [51]. Since these formulas are lengthy and complicated, it is most convenient to take the Green's functions needed in making computer programs for electromagnetic integral equations, for example, from the original publications. Of the solutions for Green's dyadics, only the two simplest, that is, the whole-space and the half-space Green's dyadics, will be given in this book. For the half-space, when both calculation point $\mathbf{R}$ and source point $\mathbf{R}_0$ are below the ground surface, a short formulation, after [52], is given in Appendix D.

6.4 VOLUME INTEGRAL EQUATIONS FOR THREE-DIMENSIONAL ELECTROMAGNETIC FIELDS

6.4.1 Electric field

In the following we shall convert the boundary-value problem defined by the wave equations (6.2.14)–(6.2.16), and boundary conditions (6.2.8)–(6.2.10) into an integral equation by applying equation (6.3.1). The treatment here essentially follows the work of Oksama [53]. Corresponding integral equations for a more restricted problem, a purely permittive and permeable model with plane-wave excitation, have been published in [54].

Let $\mathbf{R}_0$ be located, for example, in region V_1 of Fig. 6.1, and let us first apply equation (6.3.1) to region V_3. The operations are carried out with respect to variable $\mathbf{R}$. Substituting equations (6.2.16) and (6.3.3), and using the radiation conditions given by equations (6.2.10) and (6.3.6), we obtain:

$$\int_A \left[(\nabla \times \mathbf{E}_S) \times \mathbf{n} \cdot \tilde{\mathbf{G}}^{3,2} - \mathbf{n} \times \mathbf{E}_S \cdot \nabla \times \tilde{\mathbf{G}}^{3,2}\right] dA = 0, \qquad \mathbf{R} \in V_3 \quad (6.4.1)$$

Note that Green's function is entirely specified by the host structure and does not depend upon the anomalous region V_1. Hence, the second superscript remains 2 when $\mathbf{R}_0$ is in either V_1 or V_2.

Similarly, we apply equation (6.3.1) to region V_2. Substituting equations (6.2.15) and (6.3.3), we obtain:

$$\int_{S+A} \left[(\nabla \times \mathbf{E}_S) \times \mathbf{n} \cdot \tilde{\mathbf{G}}^{2,2} - \mathbf{n} \times \mathbf{E}_S \cdot \nabla \times \tilde{\mathbf{G}}^{2,2}\right] dS = 0, \qquad \mathbf{R} \in V_2 \quad (6.4.2)$$

Applying equation (6.3.1) in region V_1 and substituting equations (6.2.14) and (6.3.3), we obtain:

$$\mathbf{E}_S(\mathbf{R}_0) = \int_{V_1} \left(\Delta k^2 \mathbf{E} + \frac{\nabla \mu \times \nabla \times \mathbf{E}}{\mu}\right) \cdot \tilde{\mathbf{G}}^{2,2} \, dV +$$

$$\int_S \left[(\nabla \times \mathbf{E}_S) \times \mathbf{n} \cdot \tilde{\mathbf{G}}^{2,2} - \mathbf{n} \times \mathbf{E}_S \cdot \nabla \times \tilde{\mathbf{G}}^{2,2}\right] dS, \qquad \mathbf{R} \in V_1 \quad (6.4.3)$$

We now substitute equations (6.2.12) and boundary conditions (6.2.8), (6.2.9), (6.3.4) and (6.3.5) into equations (6.4.1) and (6.4.3). Equation (6.2.9) implies that $(1/k_2)\,\mathbf{n} \times \nabla \times \mathbf{E}_2 = (1/k_3)\,\mathbf{n} \times \nabla \times \mathbf{E}_3$.

Furthermore, by the symmetry of electric Green's dyadics, we have

$$\frac{1}{\mu_j}\,\widetilde{\overline{\overline{\mathbf{G}}}}^{j,i}(\mathbf{R}, \mathbf{R}_0) = \frac{1}{\mu_i}\,\widetilde{\mathbf{G}}^{i,j}(\mathbf{R}_0, \mathbf{R}) \qquad (6.4.4)$$

where $\widetilde{\overline{\overline{\mathbf{G}}}}$ is the transpose of $\widetilde{\mathbf{G}}$. Hence we obtain:

$$\mathbf{E}(\mathbf{R}) = \mathbf{E}_\mathrm{P}(\mathbf{R}) +$$

$$\int_{V_1} \widetilde{\mathbf{G}}^{j,2}(\mathbf{R}, \mathbf{R}_0) \cdot \left[\Delta k^2(\mathbf{R}_0)\,\mathbf{E}(\mathbf{R}_0) + \frac{\nabla_0\,\mu(\mathbf{R}_0) \times (\nabla_0 \times \mathbf{E}(\mathbf{R}_0))}{\mu(\mathbf{R}_0)} \right] dV_0$$

$$+ \int_S \widetilde{\mathbf{G}}^{j,2}(\mathbf{R}, \mathbf{R}_0) \cdot \left[\frac{(\mu_2 - \mu_1)(\mathbf{R}_0)}{\mu_1(\mathbf{R}_0)} (\nabla_0 \times \mathbf{E}(\mathbf{R}_0)) \times \mathbf{n} \right] dS_0 \qquad (6.4.5)$$

Correspondingly, the primary field $\mathbf{E}_\mathrm{P}$, defined by equations (6.2.13) and (6.2.8)–(6.2.10), reduces to the following integral form:

$$\mathbf{E}_\mathrm{P}(\mathbf{R}) = -i\omega \int_{V_S} \widetilde{\mathbf{G}}^{j,i}(\mathbf{R}, \mathbf{R}_0) \cdot \mu(\mathbf{R}_0)\,\mathbf{j}^S(\mathbf{R}_0)\,dV_0 \qquad (6.4.6)$$

where V_S is the volume containing the source currents of known density $\mathbf{j}^S$, and superscripts j and i ($j = 2, 3$; $i = 2, 3$) refer to the location regions V_j and V_i of points $\mathbf{R}$ and $\mathbf{R}_0$, respectively.

Before using equation (6.4.5) to calculate $\mathbf{E}$ in regions V_2 and V_3, the unknown $\mathbf{E}$ values involved in the integrals must first be solved in the source region by moving the calculation point $\mathbf{R}$ to V_1 and S. When this is done, the integrands in equation (6.4.5) become singular at certain points. The contribution of a singularity to the integral containing it depends upon the properties of the integral. Hence, in solving integral equations, each integral must be analytically evaluated in the vicinity of the singular points.

We see from equation (6.4.5) that the integral representation for a general model which contains anomalous distributions in each of the physical properties g, ε and μ is very complicated. The solution domain of the integral equation deducible from equation (6.4.5) includes both the anomalous volume V_1 and its surface S. The integrals in the equation contain both the unknown field strength $\mathbf{E}$ and the curl of $\mathbf{E}$. This can be expected to cause considerable difficulties in the numerical solution of the integral equation.

The anomalies obtained in electromagnetic surveys are very often caused by inhomogeneities in conductivity rather than in permeability. Hence, in order to simplify the formulas, we assume that permeability μ is the same and invariable in V_1 and V_2, in which case equation (6.4.5) simplifies to the following form:

$$E(\mathbf{R}) = E_P(\mathbf{R}) + \int_{V_1} \tilde{\mathbf{G}}^{j,2}(\mathbf{R}, \mathbf{R}_0) \cdot \Delta k^2(\mathbf{R}_0) \, \mathbf{E}(\mathbf{R}_0) \, dV_0 \qquad (6.4.7)$$

where superscript j in Green's dyadic denotes the location region of the calculation point $\mathbf{R}$. Equation (6.4.7) is similar to the well-known integral equations presented in [49; 52; 55].

It can be shown by allowing ω to approach zero and applying Gauss's law, that equation (6.4.7) reduces to equation (3.3.7), given for the static electric field in Section 3.3.2.

To solve the unknown field strength $\mathbf{E}$, a new integral equation is deduced from equation (6.4.7) by moving the calculation point $\mathbf{R}$ to the source region V_1. As can be seen from equations (6.3.11) and (6.3.13), the whole-space part $\nabla \nabla G_0$ of Green's dyadic becomes infinite as $\mathbf{R}$ approaches $\mathbf{R}_0$ and hence the integral in equation (6.4.7) is improper. The singularity is treated by removing the singular point from the integral as follows [56; 58]:

$$\mathbf{E} = \mathbf{E}_P + PV \int_{V_1} \tilde{\mathbf{G}} \cdot \Delta k^2 \, \mathbf{E} \, dV_0 - (\Delta k^2/k^2) \, \tilde{\mathbf{D}} \cdot \mathbf{E} \qquad (6.4.8)$$

where PV denotes the principal value, and

$$\tilde{\mathbf{D}} \cdot \mathbf{E} = \lim_{v \to 0} \int_v \tilde{\mathbf{G}}_0 \cdot \mathbf{E} \, dv_0 \qquad (6.4.9)$$

It can be seen from equation (6.3.12) that the field caused by the first term of $\tilde{\mathbf{G}}_0$ in $\tilde{\mathbf{D}} \cdot \mathbf{E}$ vanishes with vanishing v, and hence:

$$\tilde{\mathbf{D}} \cdot \mathbf{E} = \lim \int_v \nabla_0 \nabla_0 G_0 \cdot \mathbf{E} \, dv_0 \qquad (6.4.10)$$

Since v is very small, $\mathbf{E}$ can be considered as homogeneous in v and it can thus be taken out of the integral. Taking into consideration that G_0 approaches $1/4\pi |\mathbf{R} - \mathbf{R}_0|$, as $\mathbf{R}$ approaches $\mathbf{R}_0$, and applying Gauss's law:

$$\tilde{\mathbf{D}} = \int_v \nabla_0 \nabla_0 G_0 \, dv_0 = \int_s \mathbf{n}_0 \nabla_0 G_0 \, ds_0$$

where s is the surface bounding v, equation (6.4.10) can be written in the following form:

$$\tilde{\mathbf{D}} = \int_s \mathbf{n}_0 \nabla_0 \left(\frac{1}{4\pi |\mathbf{R} - \mathbf{R}_0|} \right) ds_0 = \int_s \frac{\mathbf{n}_0 \, \mathbf{r}_0}{4\pi} \, d\Omega_0 \qquad (6.4.11)$$

where Ω is the solid angle with origin at point $\mathbf{R}$, $\mathbf{n}_0$ is the outward unit vector of small surface s, and $\mathbf{r}_0$ is the unit vector along the vector $\mathbf{R}_0 - \mathbf{R}$.

134 Electromagnetic methods

The dyadic $\tilde{\mathbf{D}}$ may be considered as the depolarization dyadic for volume v. Its value depends upon the shape of v. Consequently, in order uniquely to specify the value of the improper integral in equation (6.4.7), the shape of the small volume v used in removing the singular point from the integral must first be specified. The volume v is usually taken to be a sphere, in which case $\tilde{\mathbf{D}} = \tilde{\mathbf{I}}/3$.

Taking the small volume enclosing the singular point $\mathbf{R} = \mathbf{R}_0$ to be a sphere, the singularity can be treated by writing the whole-space Green's dyadic of equation (6.4.7) in the following form:

$$\tilde{\mathbf{G}}_0 = \left(\tilde{\mathbf{I}} + \frac{1}{k_2^2} \nabla \nabla\right) G_0 - \frac{1}{3k_2^2} \tilde{\mathbf{I}} \, \delta(\mathbf{R} - \mathbf{R}_0) \qquad (6.4.12)$$

If in the interpretation of measurement results the magnetic field also needs to be known, it can be calculated most easily from the known electric field values using equation (6.2.1) in the following form:

$$\mathbf{H} = (i/\omega\mu) \nabla \times \mathbf{E}$$

so that

$$\mathbf{H}(\mathbf{R}) = \mathbf{H}_\mathrm{P}(\mathbf{R}) + \frac{i}{\omega\mu(\mathbf{R})} \int_{V_1} \nabla \times \tilde{\mathbf{G}}(\mathbf{R}, \mathbf{R}_0) \cdot \Delta k^2(\mathbf{R}_0) \, \mathbf{E}(\mathbf{R}_0) \, \mathrm{d}V_0 \qquad (6.4.13)$$

where

$$\mathbf{H}_\mathrm{P}(\mathbf{R}) + \frac{1}{\mu(\mathbf{R})} \int_{V_\mathrm{S}} \nabla \times \tilde{\mathbf{G}}(\mathbf{R}, \mathbf{R}_0) \cdot \mu(\mathbf{R}_0) \, \mathbf{j}^\mathrm{S}(\mathbf{R}_0) \, \mathrm{d}V_0 \qquad (6.4.14)$$

In most applications presented in the literature, integral equation (6.4.7) has been solved numerically by using the point-matching method with pulse functions as the subsectional basis (see Section 1.3.3), but higher-degree interpolation methods have also been used. Detailed descriptions of the mathematical procedures associated with the numerical solution of equation (6.4.7) are given in [50; 52; 57; 59; 60]. Homogeneous (plane-wave source) or two-dimensional (long line current source) primary fields have usually been used in these calculations, but some results for three-dimensional primary fields (small current loops) have also been presented. Numerical solutions of equation (6.4.7) are given, for example, in [50; 52; 57]. They show that with the present capacity of computers, it is possible to obtain fairly accurate numerical results for geologically relevant models.

As an application of equation (6.4.7) we consider magnetotelluric apparent resistivity anomalies at frequencies of 0.1 Hz and 10 Hz for a rectangular prism situated in a half-space [57]. The resistivity of the prism is 5 Ωm and that of the half-space 100 Ωm. The dimensions of the prism in the x, z plane are 1000 m $\times$ 2000 m, depth to the upper surface is 500 m, and the strike length parallel to the y-axis is variable. The prism and its discretization for the numerical solution are illustrated in Fig. 6.2.

Volume integral equations for 3-D fields

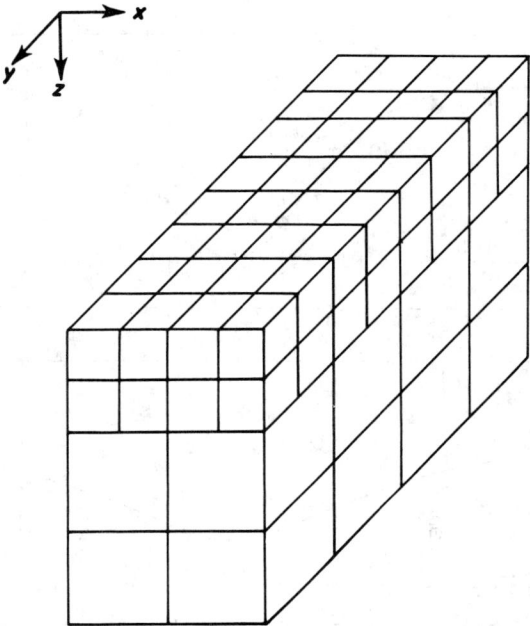

Figure 6.2 Anomalously conductive prism and scheme for discretization of its volume for numerical solution of an integral equation [57, p. 183].

The apparent resistivity (ρ_a) anomalies are calculated for two different primary field components, the $E_\parallel$ and the $E_\perp$ modes, defined on the basis of the direction of electric field **E** in relation to the strike y as follows:

$$\rho_{ayx} = (1/\omega\mu)\left|E_y/H_x\right|^2$$

$$\rho_{axy} = (1/\omega\mu)\left|E_x/H_y\right|^2$$

The magnetic field needed for calculating ρ_a is obtained by using equation (6.4.13).

Figure 6.3 illustrates the $E_\parallel$ mode (E field in the y direction) and Figure 6.4 the $E_\perp$ mode (E field in the x direction) of ρ_a for prism strike lengths of 4 km, 8 km and 12 km. Anomalies for an infinitely long prism, calculated by applying the finite element method to a completely two-dimensional model, are also shown for comparison. At the higher frequency of 10 Hz, the $E_\parallel$ mode ρ_a anomaly (Fig. 6.3) is mainly controlled by the solenoidal part of the field associated with the volume currents, the laminar part due to surface charges being less effective. At 0.1 Hz, the anomaly increases with decreasing strike length, due to the secondary electric field generated by surface charges around the ends of the prism. In the $E_\perp$ mode (Fig. 6.4), surface charges are implicitly included in both the two- and three-dimensional formulations and, therefore, the two solutions diverge less than in the $E_\parallel$ mode.

136   *Electromagnetic methods*

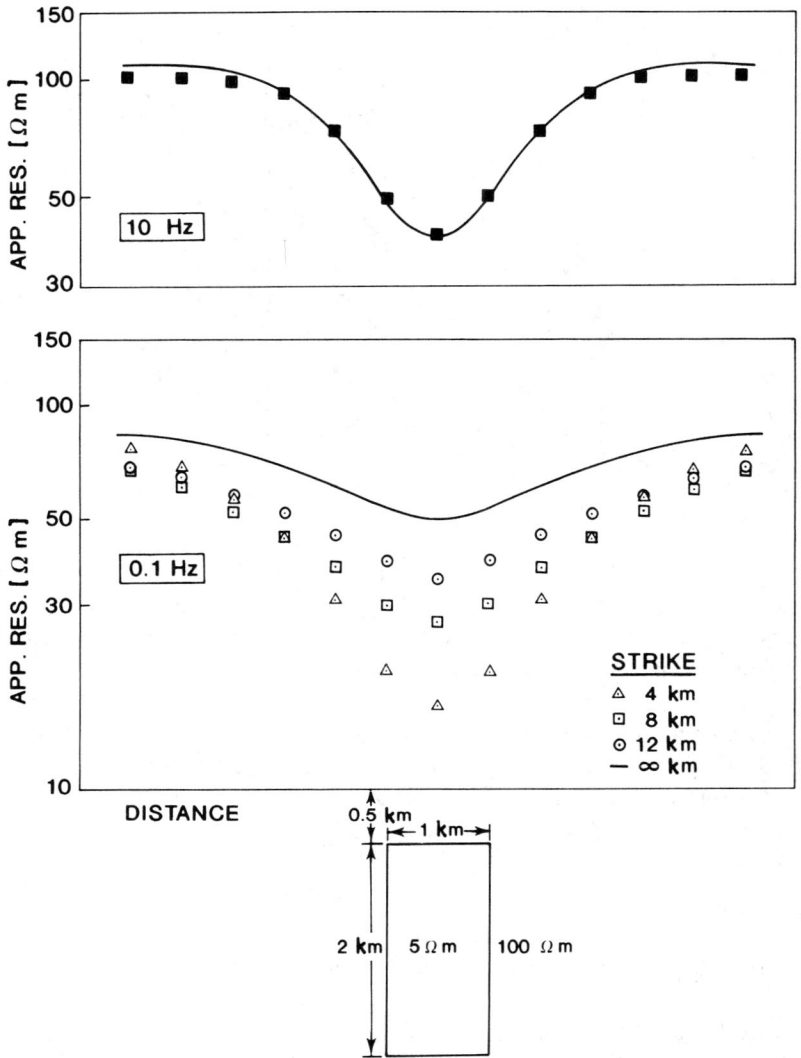

Figure 6.3 $E_\parallel$ mode of magnetotelluric apparent resistivity for a two-dimensional prism and for the three-dimensional prism of Fig. 6.2 having different strike extents [57, p. 189].

6.4.2 Magnetic field

The boundary-value problem for the field generated by magnetic sources, defined by wave equations (6.2.17)–(6.2.20) and boundary conditions (6.2.8), (6.2.9) and (6.2.11), can be converted into an integral equation by a procedure similar to that applied to the electric boundary-value problem. Substituting the

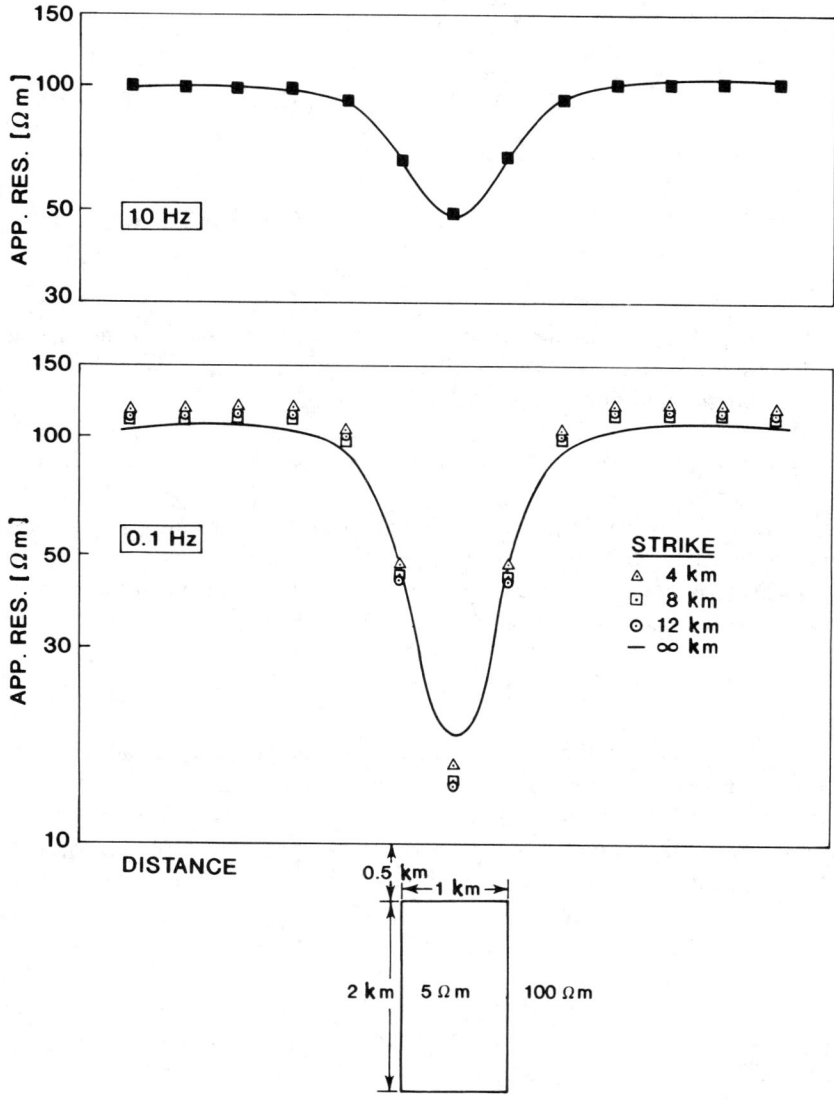

Figure 6.4 $E_\perp$ mode of magnetotelluric apparent resistivity for a two-dimensional prism and for the three-dimensional prism of Fig. 6.2 having different strike extents [57, p. 189].

secondary magnetic field $\mathbf{H}_s$ and magnetic Green's dyadic $\widetilde{\mathbf{G}}_m$ into equation (6.3.1) and applying this equation to the regions V_1, V_2 and V_3 of the earth model illustrated by Fig. 6.1, we get for the magnetic field $\mathbf{H}$ the following integral representation:

$$H(R) = H_P(R) +$$

$$\int_{V_1} \tilde{G}_m^{j,2}(R, R_0) \cdot \left[\Delta k^2(R_0) H(R_0) + \frac{\nabla_0(g + i\omega\varepsilon)(R_0)}{(g + i\omega\varepsilon)(R_0)} \times (\nabla_0 \times H(R_0))\right] dV_0$$

$$+ \int_S \tilde{G}_m^{j,2}(R, R_0) \cdot \left[\frac{((g_2 + i\omega\varepsilon_2) - (g_1 + i\omega\varepsilon_1))(R_0)}{(g_1 + i\omega\varepsilon_1)(R_0)} (\nabla_0 \times H(R_0)) \times n\right] dS_0 \quad (6.4.15)$$

The integral representation for the boundary-value problem of the primary field (defined by either equation (6.2.20) or equation (6.2.23) and the associated boundary conditions) takes the following form:

$$H_P = \int_{V_S} \tilde{G}_m^{j,i}(R, R_0) \cdot \nabla_0 \times j^S(R_0) \, dV_0 \quad (6.4.16)$$

where V_S is the volume containing the known source currents j^S, and j and i refer to the calculation and source regions V_j and V_i, respectively.

The primary field generated by magnetic sources having volume dipole moment density m^S can be obtained by conversion of equation (6.2.23):

$$H_P = \int_{V_S} \tilde{G}_m^{j,i}(R, R_0) \cdot k^2(R_0) m^S(R_0) \, dV_0 \quad (6.4.17)$$

We note again, as in the case of the corresponding electric equation (6.4.5), that for a general inhomogeneity the integral representation (6.4.15) is very complicated. Consequently, numerical applications based on this equation have not as yet appeared in the literature. However, a considerable simplification of equation (6.4.15) results if conductivity g and permittivity ε are invariable in V_1 and V_2, in which case equation (6.4.15) reduces to a volume integral:

$$H(R) = H_P(R) + \int_{V_1} \tilde{G}_m^{j,2}(R, R_0) \cdot \Delta k^2(R_0) H(R_0) \, dV_0 \quad (6.4.18)$$

where $\Delta k^2 = -i\omega(\mu_1 - \mu_2)(g + i\omega\varepsilon)$.

The electric field for the model having anomalous permeability can be obtained from the magnetic field, obtained in V_1 from equation (6.4.18), by using Maxwell's equation (6.2.2) in the following form:

$$E = (1/(g + i\omega\varepsilon))\nabla \times H$$

hence:

$$E(R) = E_P(R) + \frac{1}{(g + i\omega\varepsilon)(R)} \int_{V_1} \nabla \times \tilde{G}_m(R, R_0) \cdot \Delta k^2(R_0) H(R_0) \, dV_0 \quad (6.4.19)$$

where

$$\mathbf{E}_P(\mathbf{R}) = \frac{1}{(g + i\omega\varepsilon)(\mathbf{R})} \int_{V_S} \nabla \times \tilde{\mathbf{G}}_m(\mathbf{R}, \mathbf{R}_0) \cdot k^2(\mathbf{R}_0) \, \mathbf{m}^S(\mathbf{R}_0) \, dV_0 \quad (6.4.20)$$

It can be shown that, in the limit as ω approaches zero, equation (6.4.18) reduces to equation (4.5.7) for the static magnetic field.

6.5 VOLUME INTEGRAL EQUATIONS FOR TWO-DIMENSIONAL ELECTROMAGNETIC FIELDS

6.5.1 Introduction

A considerable saving of calculation effort results if the geological formation being surveyed is sufficiently long in one of its horizontal dimensions to be considered as a two-dimensional structural model. The solution domain, which for a three-dimensional model consists of the whole volume of the anomalous body, reduces to the cross-sectional area of the body for a two-dimensional model.

By far the simplest numerical modelling problem results when the primary field is also two-dimensional, in which case the problem can be treated directly by two-dimensional modelling. A considerably more laborious numerical solution results if the two-dimensional structural model is excited by a three-dimensional primary field, making the model $2\frac{1}{2}$-dimensional. Problems of this type are usually solved by taking the Fourier transforms of the defining functions in the elongated direction of the body. Integral equations obtained in the transform domain are then solved on the cross-sectional area of the body for a set of sufficiently sampled wavenumbers. The solution for the original problem is then obtained from the set of transform domain solutions by the inverse Fourier transformation. Solutions for the $2\frac{1}{2}$-dimensional counterpart of equation (6.4.7) have not as yet been published, whereas the completely two-dimensional solution is well documented in the literature.

6.5.2 The $E_\perp$ field

When a two-dimensional structural model is excited by a normally incident plane wave, the electromagnetic field can be separated into two independent systems, known as the $(E_\perp, H_\parallel)$ and the $(E_\parallel, H_\perp)$ modes. For a two-dimensional primary field which is homogeneous in the direction of elongation (for example, the field of a long line current source), the field can be completely represented by the $(E_\parallel, H_\perp)$ mode. Written in separated form, Maxwell's equations (6.2.1) and (6.2.2) then constitute two independent systems (see Fig. 6.5):

$$\nabla \times \mathbf{E}_\perp = -i\omega\mu \mathbf{H}_\parallel \quad (6.5.1)$$

$$\nabla \times \mathbf{H}_\parallel = (g + i\omega\varepsilon)\mathbf{E}_\perp \quad (6.5.2)$$

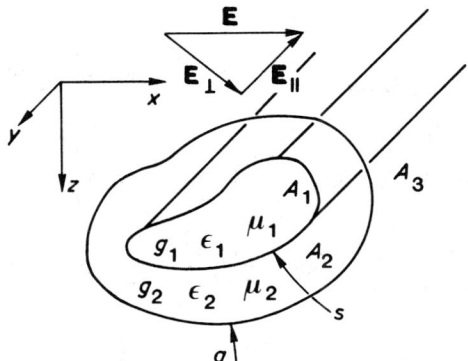

Figure 6.5 Two-dimensional structural model for an electromagnetic problem with elongation of the model parallel to the y-axis.

for the $(E_\perp, H_\parallel)$ mode, and

$$\nabla \times \mathbf{E}_\parallel = -i\omega\mu\mathbf{H}_\perp \tag{6.5.3}$$

$$\nabla \times \mathbf{H}_\perp = (g + i\omega\varepsilon)\mathbf{E}_\parallel + \mathbf{j}_\parallel^S \tag{6.5.4}$$

for the $(E_\parallel, H_\perp)$ mode.

Assuming that permeability μ is invariable in the regions illustrated by cross-sectional areas A_1 and A_2 in Fig. 6.5, we partition the field modes into primary and secondary components:

$$\mathbf{E}_\perp = \mathbf{E}_{P\perp} + \mathbf{E}_{S\perp} \tag{6.5.5}$$

$$\mathbf{E}_\parallel = \mathbf{E}_{P\parallel} + \mathbf{E}_{S\parallel} \tag{6.5.6}$$

Taking the curl on both sides in equations (6.5.1) and (6.5.2), we obtain the following wave equations for the two-dimensional electric field $\mathbf{E}$:

$$\nabla \times \nabla \times \mathbf{E}_{S\perp} - k_2^2 \mathbf{E}_{S\perp} = \Delta k^2 \mathbf{E}_\perp \tag{6.5.7}$$

when the calculation point $\mathbf{r} = x\mathbf{x} + y\mathbf{y}$ is within area A_1 of the cross-sectional plane $y = 0$, y being parallel to the elongation of the body, and $\Delta k^2 = k_1^2 - k_2^2$. Outside surface s (Fig. 6.5) $\Delta k^2 = 0$. Consequently,

$$\nabla \times \nabla \times \mathbf{E}_{S\perp} - k_2^2 \mathbf{E}_{S\perp} = 0, \quad \mathbf{r} \in A_2 \tag{6.5.8}$$

$$\nabla \times \nabla \times \mathbf{E}_{S\perp} - k_3^2 \mathbf{E}_{S\perp} = 0, \quad \mathbf{r} \in A_3 \tag{6.5.9}$$

Assuming that the source currents of the incident plane wave are outside surface a (Fig. 6.5) the primary component of $\mathbf{E}_\perp$ satisfies the wave equation:

$$\nabla \times \nabla \times \mathbf{E}_{P\perp} - k_j^2 \mathbf{E}_{P\perp} = 0 \tag{6.5.10}$$

where $j = 2$ when $\mathbf{r}$ is either within area A_1 or A_2, and $j = 3$ when $\mathbf{r}$ is on A_3.
Boundary conditions (6.2.8) and (6.2.9) for $\mathbf{E}_\perp$ take the following form:

$$\mathbf{t} \cdot \mathbf{E}_{\perp j} = \mathbf{t} \cdot \mathbf{E}_{\perp j+1} \tag{6.5.11}$$

$$\frac{1}{\mu_j} \mathbf{y} \cdot \nabla \times \mathbf{E}_{\perp j} = \frac{1}{\mu_{j+1}} \mathbf{y} \cdot \nabla \times \mathbf{E}_{\perp j+1} \tag{6.5.12}$$

along curves s ($j = 1$) and a ($j = 2$), with $\mathbf{t}$ standing for the unit tangential vector pointing around s and a in an anticlockwise sense.

Green's dyadic $\overline{\mathbf{G}}_\perp$ for the boundary-value problem under consideration operates on source and field vectors that are both aligned on the plane $y = 0$ and, hence, satisfies the following wave equation:

$$\nabla \times \nabla \times \overline{\mathbf{G}}_\perp(\mathbf{r}, \mathbf{r}_0) - k^2 \overline{\mathbf{G}}_\perp(\mathbf{r}, \mathbf{r}_0) = (\mathbf{x}\mathbf{x} + \mathbf{z}\mathbf{z}) \delta(\mathbf{r} - \mathbf{r}_0) \tag{6.5.13}$$

with points $\mathbf{r}$ and $\mathbf{r}_0$ being on plane $y = 0$ and $\partial/\partial y = 0$. The propagation constant k, defined by $k^2 = -i\omega\mu(g + i\omega\varepsilon)$, takes the physical properties of A_2 if $\mathbf{r}$ is in either A_1 or A_2 or of A_3, if $\mathbf{r}$ is in A_3. The solution of equation (6.5.13) must satisfy the following boundary conditions:

$$\mathbf{t} \cdot \overline{\mathbf{G}}_{\perp 2} = \mathbf{t} \cdot \overline{\mathbf{G}}_{\perp 3} \tag{6.5.14}$$

$$\frac{1}{\mu_2} \mathbf{y} \cdot \nabla \times \overline{\mathbf{G}}_{\perp 2} = \frac{1}{\mu_3} \mathbf{y} \cdot \nabla \times \overline{\mathbf{G}}_{\perp 3} \tag{6.5.15}$$

on curve a.

The solution for equations (6.5.13)–(6.5.15) can be written in component form as follows:

$$\overline{\mathbf{G}} = \overline{g}_{xx} \mathbf{x}\mathbf{x} + \overline{g}_{xz} \mathbf{z}\mathbf{x} + \overline{g}_{zx} \mathbf{x}\mathbf{z} + \overline{g}_{zz} \mathbf{z}\mathbf{z} \tag{6.5.16}$$

where the relation of components $\overline{g}_{ij}$ ($i = x, z$ and $j = x, z$) to the corresponding components of the three-dimensional dyadic represented by equation (6.3.2), can be written as follows:

$$\overline{g}_{ij}(\mathbf{r}, \mathbf{r}_0) = \int_{-\infty}^{\infty} g_{ij}(\mathbf{R}, \mathbf{R}_0) \, dy_0 \tag{6.5.17}$$

with the point $\mathbf{R}$ on the plane $y = 0$

The boundary-value problem defined by equations (6.5.7)–(6.5.15) can be converted into an integral equation by applying equation (6.3.1) just as in Section 6.4 for the three-dimensional model. We shall give the resulting integral formulas directly, without repeating the conversion procedure. The same integral equations are given by Lee and Morrison [61].

As a result of conversion, the following two-dimensional integral equation over an anomalous area A_1 is obtained for $\mathbf{E}_\perp$:

$$\mathbf{E}_\perp(\mathbf{r}) = \mathbf{E}_{P\perp}(\mathbf{r}) + \int_{A_1} \overline{\mathbf{G}}_\perp(\mathbf{r}, \mathbf{r}_0) \cdot \Delta k^2(\mathbf{r}_0) \, \mathbf{E}_\perp(\mathbf{r}_0) \, dA_0 \tag{6.5.18}$$

where the primary field is that of a normally incident plane wave, hence $\mathbf{E}_{P\perp}(\mathbf{r}) = \mathbf{E}_{P\perp}(z)$. Equation (6.5.18) can be solved for $\mathbf{E}_\perp$ in A_1 using standard numerical methods, by moving the point $\mathbf{r}$ to area A_1. After $\mathbf{E}_\perp$ is obtained for any point in A_1, equation (6.9.18) can be used to calculating $\mathbf{E}_\perp$ for any arbitrary point of the model.

The magnetic field $\mathbf{H}_\parallel$, associated with the electric field $\mathbf{E}_\perp$, can be obtained from equation (6.5.18) by applying equation (6.5.1), hence:

$$\mathbf{H}_\parallel(\mathbf{r}) = \mathbf{H}_{P\parallel}(\mathbf{r}) + \frac{i}{\omega\mu(\mathbf{r})} \int_{A_1} \nabla \times \overline{\mathbf{G}}_\perp(\mathbf{r}, \mathbf{r}_0) \cdot \Delta k^2(\mathbf{r}_0) \, \mathbf{E}_\perp(\mathbf{r}_0) \, dA_0 \quad (6.5.19)$$

where

$$\mathbf{H}_{P\parallel}(\mathbf{r}) = \frac{i}{\omega\mu(\mathbf{r})} \nabla \times \mathbf{E}_{P\perp}(\mathbf{r}) \quad (6.5.20)$$

A detailed description of the solution of Green's dyadic $\overline{\mathbf{G}}_\perp$ and integrations involved in the numerical solution of equations (6.5.18) and (6.5.19) is given by Lee and Morrison [62]. The elements of $\overline{\mathbf{G}}_\perp$ for a half-space can be expressed analytically, while for more complicated host structures, numerical integrations are required.

6.5.3 The $\mathbf{E}_\parallel$ field

The boundary-value problem for $\mathbf{E}_\parallel$ can be formulated by following a procedure similar to that used in formulating the boundary-value problem for $\mathbf{E}_\perp$. Taking the curl on both sides of equations (6.5.3) and (6.5.4) and eliminating the magnetic field, we obtain:

$$\nabla \times \nabla \times \mathbf{E}_{S\parallel} - k_2^2 \mathbf{E}_{S\parallel} = \Delta k^2 \mathbf{E}_\parallel, \quad \mathbf{r} \in A_1 \quad (6.5.21)$$

$$\nabla \times \nabla \times \mathbf{E}_{S\parallel} - k_2^2 \mathbf{E}_{S\parallel} = 0, \quad \mathbf{r} \in A_2 \quad (6.5.22)$$

$$\nabla \times \nabla \times \mathbf{E}_{S\parallel} - k_3^2 \mathbf{E}_{S\parallel} = 0, \quad \mathbf{r} \in A_3 \quad (6.5.23)$$

The wave equation for the primary component of $\mathbf{E}_\parallel$ is

$$\nabla \times \nabla \times \mathbf{E}_{P\parallel} - k_j^2 \mathbf{E}_{P\parallel} = -i\omega\mu \mathbf{j}_\parallel^S \quad (6.5.24)$$

where $\mathbf{j}_\parallel^S$ is the density of the source current flowing in the model. For the case of a plane wave excitation, $\mathbf{j}_\parallel^S = 0$.

Taking the fixed direction of $\mathbf{E}_\parallel$ into consideration, the boundary conditions (6.2.8) and (6.2.9), when applied to $\mathbf{E}_\parallel$, can be written in the following form:

$$\mathbf{E}_{\parallel j} = \mathbf{E}_{\parallel j+1} \quad (6.5.25)$$

$$\frac{1}{\mu_j} \mathbf{t} \cdot \nabla \times \mathbf{E}_{\parallel j} = \frac{1}{\mu_{j+1}} \mathbf{t} \cdot \nabla \times \mathbf{E}_{\parallel j+1} \quad (6.5.26)$$

Volume integral equations for 2-D fields

along curves s ($j = 1$) and a ($j = 2$).

Green's dyadic $\overline{\mathbf{G}}_{\parallel}$ for the $\mathbf{E}_{\parallel}$ field, which is aligned parallel to the y-axis, satisfies the non-homogeneous wave equation:

$$\nabla \times \nabla \times \overline{\mathbf{G}}_{\parallel}(\mathbf{r}, \mathbf{r}_0) - k^2 \, \overline{\mathbf{G}}_{\parallel}(\mathbf{r}, \mathbf{r}_0) = \mathbf{y}\mathbf{y} \, \delta(\mathbf{r} - \mathbf{r}_0) \qquad (6.5.27)$$

with $\partial/\partial y = 0$. The propagation constant k takes the same physical properties as in equation (6.5.13). The solution of equation (6.5.27) must satisfy the following boundary conditions:

$$\overline{\mathbf{G}}_{\parallel 2} = \overline{\mathbf{G}}_{\parallel 3} \qquad (6.5.28)$$

$$\frac{1}{\mu_2} \mathbf{t} \cdot \nabla \times \overline{\mathbf{G}}_{\parallel 2} = \frac{1}{\mu_3} \mathbf{t} \cdot \nabla \times \overline{\mathbf{G}}_{\parallel 3} \qquad (6.5.29)$$

along curve a.

The solution of (6.5.27)–(6.5.29) consists of only one component, $\overline{\mathbf{G}}_{\parallel} = \overline{g}_{yy} \, \mathbf{y}\mathbf{y}$, and can thus be expressed by a single scalar function $\overline{g}_{\parallel}$, which is related to the corresponding component g_{yy} of the three-dimensional dyadic of equation (6.3.2) by the integral:

$$\overline{g}_{\parallel}(\mathbf{r}, \mathbf{r}_0) = \int_{-\infty}^{\infty} g_{yy}(\mathbf{R}, \mathbf{R}_0) \, dy_0 \qquad (6.5.30)$$

where $\mathbf{R}$ is on the plane $y = 0$.

As a simple example of two-dimensional Green's dyadics we give here, after [59], $\overline{g}_{\parallel}$ for the half-space below the ground surface. Expressing $\overline{g}_{\parallel}$ as the sum of the whole-space and the secondary components, $\overline{g}_{\parallel} = \overline{g}_0 + \overline{g}_s$, we obtain for the dyadic:

$$\overline{g}_0(\mathbf{r}, \mathbf{r}_0) = -\frac{i\omega\mu}{2\pi} K_0(ik|\mathbf{r} - \mathbf{r}_0|) \qquad (6.5.31)$$

$$\overline{g}_s(\mathbf{r}, \mathbf{r}_0) = -\frac{i\omega\mu}{2\pi} \int_0^{\infty} \frac{\mu - \lambda}{u(u + \lambda)} e^{-u(z + z_0)} \cos[\lambda(x - x_0)] \, d\lambda \qquad (6.5.32)$$

with $u = (\lambda^2 - k^2)^{\frac{1}{2}}$. K_0 is the modified Bessel function of the second kind of order zero, and the physical properties involved in k are those of the earth medium.

Conversion of the boundary-value problem defined by equations (6.5.21)–(6.5.29) by applying the vector-dyadic Green's second identity (equation (6.3.1) yields for $\mathbf{E}_{\parallel}$:

$$\mathbf{E}_{\parallel}(\mathbf{r}) = \mathbf{E}_{P\parallel}(\mathbf{r}) + \int_{A_1} \overline{\mathbf{G}}_{\parallel}(\mathbf{r}, \mathbf{r}_0) \cdot \Delta k^2(\mathbf{r}_0) \, \mathbf{E}_{\parallel}(\mathbf{r}_0) \, dA_0 \qquad (6.5.33)$$

Equation (6.5.33) is similar to that given by Hohmann [63].

144   *Electromagnetic methods*

The primary field $\mathbf{E}_{P\parallel}$ can be either a normally incident plane wave or a two-dimensional field due to a two-dimensional current distribution flowing parallel to the elongation direction of the target. For the latter case

$$\mathbf{E}_{P\parallel} = -i\omega \int_{A_S} \overline{\mathbf{G}}_{\parallel}(\mathbf{r}, \mathbf{r}_0) \cdot \mu(\mathbf{r}_0) \, \mathbf{j}_{\parallel}^S(\mathbf{r}_0) \, dA_0 \qquad (6.5.34)$$

where A_S is the area containing the source current. When the current distribution is a long line current of strength $I_\parallel$ concentrated at $\mathbf{r}_S$, then

$$\mathbf{E}_{P\parallel} = -i\omega \, \overline{\mathbf{G}}_{\parallel}(\mathbf{r}, \mathbf{r}_S) \cdot \mu(\mathbf{r}_S) \, \mathbf{I}_{\parallel}(\mathbf{r}_S) \qquad (6.5.35)$$

After numerical solution of equation (6.5.33) in A_1, the electric field $\mathbf{E}_\parallel$ can be calculated at any arbitrary point $\mathbf{r}$ of the model.

The magnetic field $\mathbf{H}_\perp$ can be calculated by using equation (6.5.3), giving

$$\mathbf{H}_\perp(\mathbf{r}) = \mathbf{H}_{P\perp}(\mathbf{r}) + \frac{i}{\omega\mu(\mathbf{r})} \int_{A_1} \nabla \times \overline{\mathbf{G}}_{\parallel}(\mathbf{r}, \mathbf{r}_0) \cdot \Delta k^2(\mathbf{r}_0) \, \mathbf{E}_\parallel(\mathbf{r}_0) \, dA_0 \qquad (6.5.36)$$

with

$$\mathbf{H}_{P\perp}(\mathbf{r}) = \frac{i}{\omega\mu(\mathbf{r})} \nabla \times \mathbf{E}_{P\parallel}(\mathbf{r}) \qquad (6.5.37)$$

Detailed descriptions of subsectional discretization of the model and integrations of Green's functions associated with the numerical solution of equation (6.5.31) are given, for example, by Hohmann [59; 63].

6.6 SURFACE INTEGRAL EQUATIONS FOR ELECTROMAGNETIC FIELDS

When the physical properties in the anomalous region of an earth model are homogeneous, an electromagnetic boundary-value problem in the model can be solved by converting it into a set of surface integral equations. Surface integral equations for electric and magnetic fields can be formulated by applying the vector-dyadic Green's second identity (6.3.1), just as for the general integral equations (6.4.5) and (6.4.15). By performing different kinds of mathematical operation in the application of equation (6.3.1), different types of surface integral equation are obtained, of which the classical equations of Stratton and Chu [6], and Muller [45] may be mentioned. In this book, we shall only deal with the formulations by Doherty [44] for solving electromagnetic boundary-value problems by means of a set of coupled surface integral equations. The secondary sources for the electric and magnetic fields are represented by equivalent tangential electric and magnetic sources distributed on the surface of the anomalous body. The surface integral equations thus obtained resemble very closely those given by Muller. Among the surface integral equations to be found in the literature, Doherty's equations are probably those which are best suited to numerical solution.

6.6.1 Three-dimensional model

Let the boundary-value problem for the electric and magnetic fields **E** and **H** in the structural model illustrated by Fig. 6.1 be defined, in accordance with [44], within the regions V_1, V_2 and V_3, by Maxwell's equations:

$$\nabla \times \mathbf{E} = -i\omega\mu\mathbf{H} \qquad (6.6.1)$$

$$\nabla \times \mathbf{H} = (g + i\omega\varepsilon)\mathbf{E} \qquad (6.6.2)$$

The solution to these equations must satisfy the radiation conditions

$$\mathbf{E} \times \frac{\mathbf{R}}{R} + \left(\frac{i\omega\mu}{(g + i\omega\varepsilon)}\right)\mathbf{H} = 0\left(\frac{1}{R}\right) \qquad (6.6.3)$$

$$\mathbf{H} \times \frac{\mathbf{R}}{R} - \left(\frac{(g + i\omega\varepsilon)}{i\omega\mu}\right)\mathbf{E} = 0\left(\frac{1}{R}\right) \qquad (6.6.4)$$

as R approaches infinity in region V_3, and the boundary conditions

$$\mathbf{n} \times \mathbf{E}_j = \mathbf{n} \times \mathbf{E}_{j+1} \qquad (6.6.5)$$

$$\mathbf{n} \times \mathbf{H}_j = \mathbf{n} \times \mathbf{H}_{j+1} \qquad (6.6.6)$$

at surfaces S ($j=1$) and A ($j=2$). Note that in [44], the time dependence of the fields is considered to be $\exp(-i\omega t)$ whereas, in this book, the time dependence $\exp(i\omega t)$ is used. The formulas given here are obtained from those of [44] by substituting $i\omega$ for $-i\omega$.

Let the electric and magnetic fields be partitioned into primary and secondary components, $\mathbf{E} = \mathbf{E}_P + \mathbf{E}_S$ and $\mathbf{H} = \mathbf{H}_P + \mathbf{H}_S$, and let us assume that the primary source currents are outside the anomalous region V_1. The total fields **E** and **H** in region V_1, and the secondary fields $\mathbf{E}_S$, $\mathbf{H}_S$ in regions V_2 and V_3 are considered to be generated by tangential electric and magnetic source densities **a** and **b** distributed on surface S. The electric and magnetic fields can then be expressed in terms of the convolutions of suitable Green's dyadics and the source functions **a** and **b**. The electric and magnetic Green's dyadics of the problem consist of two partial dyadics as follows:

$$\tilde{\mathbf{G}}^E = \begin{pmatrix} \mathbf{G}^{ee} \\ \mathbf{G}^{em} \end{pmatrix}$$

$$\tilde{\mathbf{G}}^M = \begin{pmatrix} \mathbf{G}^{me} \\ \mathbf{G}^{mm} \end{pmatrix} \qquad (6.6.7)$$

where $\mathbf{G}^{ee}$ and $\mathbf{G}^{em}$ involve the electric field due to unit electric and magnetic sources, and $\mathbf{G}^{me}$ and $\mathbf{G}^{mm}$ involve the magnetic field due to these sources. The dyadics $\mathbf{G}^{ee}$ and $\mathbf{G}^{mm}$ have already been considered in this book in Section 6.4, where they were called $\widetilde{\mathbf{G}}$ and $\widetilde{\mathbf{G}}_m$, respectively.

Let us partition the partial dyadics into whole-space and secondary components as follows:

$$\mathbf{G}^{ee} = \mathbf{G}_0^{ee} + \mathbf{G}_S^{ee}$$

$$\mathbf{G}^{em} = \mathbf{G}_0^{em} + \mathbf{G}_S^{em}$$

$$\mathbf{G}^{me} = \mathbf{G}_0^{me} + \mathbf{G}_S^{me}$$

$$\mathbf{G}^{mm} = \mathbf{G}_0^{mm} + \mathbf{G}_S^{mm} \tag{6.6.8}$$

Using this notation, the electric and magnetic fields in regions V_1 and V_2 can be represented by surface integrals as follows:

$$\mathbf{E}_1(\mathbf{R}) = \int_S \{\mathbf{G}_{01}^{em}(\mathbf{R}, \mathbf{R}_0) \cdot \mathbf{b}(\mathbf{R}_0) + \mathbf{G}_{01}^{ee}(\mathbf{R}, \mathbf{R}_0) \cdot \mathbf{a}(\mathbf{R}_0)\} \, dS_0 \tag{6.6.9}$$

$$\mathbf{H}_1(\mathbf{R}) = \int_S \{\mathbf{G}_{01}^{me}(\mathbf{R}, \mathbf{R}_0) \cdot \mathbf{a}(\mathbf{R}_0) + \mathbf{G}_{01}^{mm}(\mathbf{R}, \mathbf{R}_0) \cdot \mathbf{b}(\mathbf{R}_0)\} \, dS_0 \tag{6.6.10}$$

in anomalous region V_1, and

$$\mathbf{E}_2(\mathbf{R}) = \mathbf{E}_P(\mathbf{R}) + \int_S \{[\mathbf{G}_{02}^{em}(\mathbf{R}, \mathbf{R}_0) + \mathbf{G}_S^{em}(\mathbf{R}, \mathbf{R}_0)] \cdot \mathbf{b}(\mathbf{R}_0)$$
$$+ [\mathbf{G}_{02}^{ee}(\mathbf{R}, \mathbf{R}_0) + \mathbf{G}_S^{ee}(\mathbf{R}, \mathbf{R}_0)] \cdot \mathbf{a}(\mathbf{R}_0)\} \, dS_0 \tag{6.6.11}$$

$$\mathbf{H}_2(\mathbf{R}) = \mathbf{H}_P(\mathbf{R}) + \int_S \{[\mathbf{G}_{02}^{me}(\mathbf{R}, \mathbf{R}_0) + \mathbf{G}_S^{me}(\mathbf{R}, \mathbf{R}_0)] \cdot \mathbf{a}(\mathbf{R}_0)$$
$$+ [\mathbf{G}_{02}^{mm}(\mathbf{R}, \mathbf{R}_0) + \mathbf{G}_S^{mm}(\mathbf{R}, \mathbf{R}_0)] \cdot \mathbf{b}(\mathbf{R}_0)\} \, dS_0 \tag{6.6.12}$$

in region V_2.

The operations under the integrals by the whole-space Green's dyadic components are carried out in [44] as follows:

$$\mathbf{G}_{0j}^{em}(\mathbf{R}, \mathbf{R}_0) \cdot = -i\omega\mu_j \nabla \times G_{0j}(\mathbf{R}, \mathbf{R}_0) \tag{6.6.13}$$

$$\mathbf{G}_{0j}^{ee}(\mathbf{R}, \mathbf{R}_0) \cdot = \nabla \times \nabla \times G_{0j}(\mathbf{R}, \mathbf{R}_0) \tag{6.6.14}$$

$$\mathbf{G}_{0j}^{me}(\mathbf{R}, \mathbf{R}_0) \cdot = (g_j + i\omega\varepsilon_j)\nabla \times G_{0j}(\mathbf{R}, \mathbf{R}_0) \tag{6.6.15}$$

$$\mathbf{G}_{0j}^{mm}(\mathbf{R}, \mathbf{R}_0) \cdot = \nabla \times \nabla \times G_{0j}(\mathbf{R}, \mathbf{R}_0) \tag{6.6.16}$$

where j (= 1, 2) refers to regions V_1 and V_2.

G_{0j} is the singular whole-space Green's function, considered in Section 6.3:

$$G_{0j}(\mathbf{R}, \mathbf{R}_0) = \frac{e^{-ik_j|\mathbf{R}-\mathbf{R}_0|}}{4\pi|\mathbf{R}-\mathbf{R}_0|} \qquad (6.6.17)$$

where k_j, defined by $k_j^2 = -i\omega\mu_j(g_j + i\omega\varepsilon_j)$, takes the physical properties of region V_j ($j = 1, 2$). The dyadics $\mathbf{G}_S$, operating in V_2 (and in V_3), are the non-singular secondary Green's dyadics representing the fields due to electromagnetic scattering on boundary A (see Fig. 6.1).

The secondary dyadics $\mathbf{G}_S$ can be solved by following the principles considered in Section 6.3. Note that the effects in region V_1 due to the primary sources and the scattering on boundary A, are incorporated in the source densities $\mathbf{a}$ and $\mathbf{b}$ distributed on S.

The fields $\mathbf{E}_1$, $\mathbf{H}_1$ and $\mathbf{E}_2$, $\mathbf{H}_2$ represented by equations (6.6.9)–(6.6.16) satisfy Maxwell's equations within their domains V_1 and V_2.

To solve the unknown source densities $\mathbf{a}$ and $\mathbf{b}$ on surface S, the boundary conditions related to $\mathbf{E}$ and $\mathbf{H}$ are enforced by substituting equations (6.6.9)–(6.6.16) into equations (6.6.5) and (6.6.6). Taking into consideration the effect of discontinuity in the integration of the $\nabla\times$ term as the point $\mathbf{R}$ is moved on to S in each of equations (6.6.13)–(6.6.16), the following Fredholm surface integral equation system of the second kind is obtained:

$$\frac{-i\omega(\mu_2+\mu_1)\mathbf{b}(\mathbf{R})}{2}$$

$$\int_S \mathbf{n} \times \{\nabla \times [i\omega\mu_2 G_{02}(\mathbf{R},\mathbf{R}_0) - i\omega\mu_1 G_{01}(\mathbf{R},\mathbf{R}_0)]\mathbf{b}(\mathbf{R}_0) + \mathbf{G}_S^{em}(\mathbf{R},\mathbf{R}_0)\cdot\mathbf{b}(\mathbf{R}_0)\}\,dS_0 +$$

$$\int_S \mathbf{n} \times \{\nabla \times \nabla \times [G_{02}(\mathbf{R},\mathbf{R}_0) - G_{01}(\mathbf{R},\mathbf{R}_0)]\mathbf{a}(\mathbf{R}_0) + \mathbf{G}_S^{ee}(\mathbf{R},\mathbf{R}_0)\cdot\mathbf{a}(\mathbf{R}_0)\}\,dS_0$$

$$= -\mathbf{n} \times \mathbf{E}_P(\mathbf{R}) \qquad (6.6.18)$$

$$\frac{[(g_2+i\omega\varepsilon_2) + (g_1+i\omega\varepsilon_1)]\mathbf{a}(\mathbf{R})}{2}$$

$$+ \int_S \mathbf{n} \times \{\nabla \times [(g_2+i\omega\varepsilon_2)G_{02}(\mathbf{R},\mathbf{R}_0) - (g_1+i\omega\varepsilon_1)G_{01}(\mathbf{R},\mathbf{R}_0)]\mathbf{a}(\mathbf{R}_0)$$

$$+ \mathbf{G}_S^{me}(\mathbf{R},\mathbf{R}_0)\cdot\mathbf{a}(\mathbf{R}_0)\}\,dS_0$$

$$+ \int_S \mathbf{n} \times \{\nabla \times \nabla \times [G_{02}(\mathbf{R},\mathbf{R}_0) - G_{01}(\mathbf{R},\mathbf{R}_0)]\mathbf{b}(\mathbf{R}_0) + \mathbf{G}_S^{mm}(\mathbf{R},\mathbf{R}_0)\cdot\mathbf{b}(\mathbf{R}_0)\}\,dS_0$$

$$= -\mathbf{n} \times \mathbf{H}_P(\mathbf{R}) \qquad (6.6.19)$$

148 *Electromagnetic methods*

when **R** is on surface S; **n** stands for the outward unit normal vector of S. The improper integrals are taken as the principal values.

After numerical solution of the set of integral equations (6.6.18) and (6.6.19) for **a** and **b**, the electric and magnetic fields **E** and **H** can be calculated at any arbitrary point of the model by using equations (6.6.9)–(6.6.16).

6.6.2 $2\tfrac{1}{2}$-dimensional model

A considerable simplification in the numerical solution of a boundary-value problem is achieved when the model is two-dimensional. If a two-dimensional structural model is excited by a three-dimensional primary field, then the resulting boundary-value problem may be most efficiently treated by a $2\tfrac{1}{2}$-dimensional modelling technique. In $2\tfrac{1}{2}$-dimensional modelling, Fourier transforms are calculated for the functions forming the boundary-value problem, the transforms being taken in the direction of elongation of the model. The transformed boundary-value problem is then solved for a properly sampled set of wavenumbers. The solution for the original space domain problem can then be obtained from the set of transform domain solutions by the inverse Fourier transformation.

Surface integral equations for a completely two-dimensional model were given, for example, by Parry and Ward [64] as early as 1971. Of the surface integral equations available for electromagnetic problems in two-dimensional structural models, we shall present here the $2\tfrac{1}{2}$-dimensional solution by Doherty, which is essentially an application to a two-dimensional structural model of the solution considered in the previous section.

Let the boundary-value problem again be given by equations (6.6.1)–(6.6.6), but assume now that the structural model is two-dimensional, as illustrated in Fig. 6.5, the elongated dimension of the model being parallel to the y-axis of a Cartesian system of coordinates x, y, z. Let the Fourier transform pair relative to y be defined as:

$$F_y\{f(\mathbf{R})\} = \overline{f}(\mathbf{r}, k_y) = \int_{-\infty}^{\infty} f(\mathbf{R})\, e^{-ik_y y}\, dy \qquad (6.6.20a)$$

$$F_y^{-1}\{\overline{f}(\mathbf{r}, k_y)\} = f(\mathbf{R}) = \frac{1}{2\pi} \int_{-\infty}^{\infty} \overline{f}(\mathbf{r}, k_y)\, e^{ik_y y}\, dk_y \qquad (6.6.20b)$$

where $\mathbf{R} = x\mathbf{x} + y\mathbf{y} + z\mathbf{z}$ is a three-dimensional position vector with **x**, **y** and **z** being the unit vectors along the coordinate axes x, y and z, $\mathbf{r} = x\mathbf{x} + z\mathbf{z}$ is the two-dimensional projection of **R** on to plane $y = 0$, and k_y is the wavenumber in the y direction.

Let **n** be the outward unit normal vector of surface S, **t** the unit tangential vector of cross-sectional contour s of cylindrical surface S in an anticlockwise sense (see Fig. 6.5). Taking Fourier transforms (6.6.20a) of equations (6.6.5) and (6.6.6), we obtain the boundary conditions satisfied by the y and t components of the electric and magnetic fields $\overline{\mathbf{E}}$ and $\overline{\mathbf{H}}$ in the transform domain:

$$\overline{E}_{yj} = \overline{E}_{yj+1} \qquad (6.6.21)$$

$$\mathbf{t} \cdot \overline{\mathbf{E}}_j = \mathbf{t} \cdot \overline{\mathbf{E}}_{j+1} \qquad (6.6.22)$$

$$\overline{H}_{yj} = \overline{H}_{yj+1} \qquad (6.6.23)$$

$$\mathbf{t} \cdot \overline{\mathbf{H}}_j = \mathbf{t} \cdot \overline{\mathbf{H}}_{j+1} \qquad (6.6.24)$$

on contours s ($j=1$) and a ($j=2$). $\overline{E}_y$ and $\overline{H}_y$ stand for the y components of the transformed fields $\overline{\mathbf{E}}$ and $\overline{\mathbf{H}}$. Following [44], the total electric and magnetic fields $\overline{\mathbf{E}}$ and $\overline{\mathbf{H}}$ in the anomalous region V_1, and the secondary fields $\overline{\mathbf{E}}_S$ and $\overline{\mathbf{H}}_S$ outside V_1 are considered to be generated by tangential electric and magnetic source densities $\mathbf{a}$ and $\mathbf{b}$, distributed on the cylindrical surface S of body V_1. Let the Fourier transforms of $\mathbf{a}$ and $\mathbf{b}$ relative to y be denoted $\overline{\mathbf{a}}$ and $\overline{\mathbf{b}}$, and let the y and t components of $\overline{\mathbf{a}}$ and $\overline{\mathbf{b}}$ be $\overline{a}_y$, $\overline{a}_t$ and $\overline{b}_y$, $\overline{b}_t$, respectively. Taking Fourier transforms of equations (6.6.9)–(6.6.16), we obtain the following line integral representations for the electric and magnetic fields in the transform domain:

$$\overline{\mathbf{E}}_1(\mathbf{r}) = \int_S \left\{ \overline{\mathbf{G}}_{01}^{ee}(\mathbf{r}, \mathbf{r}_0) \cdot \overline{\mathbf{a}}(\mathbf{r}_0) + \overline{\mathbf{G}}_{01}^{em}(\mathbf{r}, \mathbf{r}_0) \cdot \overline{\mathbf{b}}(\mathbf{r}_0) \right\} ds_0 \qquad (6.6.25)$$

$$\overline{\mathbf{H}}_1(\mathbf{r}) = \int_S \left\{ \overline{\mathbf{G}}_{01}^{me}(\mathbf{r}, \mathbf{r}_0) \cdot \overline{\mathbf{a}}(\mathbf{r}_0) + \overline{\mathbf{G}}_{01}^{mm}(\mathbf{r}, \mathbf{r}_0) \cdot \overline{\mathbf{b}}(\mathbf{r}_0) \right\} ds_0 \qquad (6.6.26)$$

in the anomalous area A_1 of plane $y=0$, and

$$\overline{\mathbf{E}}_2(\mathbf{r}) = \overline{\mathbf{E}}_P(\mathbf{r}) + \int_S \left\{ \left[\overline{\mathbf{G}}_{02}^{ee}(\mathbf{r}, \mathbf{r}_0) + \overline{\mathbf{G}}_S^{ee}(\mathbf{r}, \mathbf{r}_0) \right] \cdot \overline{\mathbf{a}}(\mathbf{r}_0) \right.$$
$$\left. + \left[\overline{\mathbf{G}}_{02}^{em}(\mathbf{r}, \mathbf{r}_0) + \overline{\mathbf{G}}_S^{em}(\mathbf{r}, \mathbf{r}_0) \right] \cdot \overline{\mathbf{b}}(\mathbf{r}_0) \right\} ds_0 \qquad (6.6.27)$$

$$\overline{\mathbf{H}}_2(\mathbf{r}) = \mathbf{H}_P(\mathbf{r}) + \int_S \left\{ \left[\overline{\mathbf{G}}_{02}^{me}(\mathbf{r}, \mathbf{r}_0) + \overline{\mathbf{G}}_S^{me}(\mathbf{r}, \mathbf{r}_0) \right] \cdot \overline{\mathbf{a}}(\mathbf{r}_0) \right.$$
$$\left. + \left[\overline{\mathbf{G}}_{02}^{mm}(\mathbf{r}, \mathbf{r}_0) + \overline{\mathbf{G}}_S^{mm}(\mathbf{r}, \mathbf{r}_0) \right] \cdot \overline{\mathbf{b}}(\mathbf{r}_0) \right\} ds_0 \qquad (6.6.28)$$

in area A_2.

Green's dyadics in the transform domain are obtained by taking the Fourier transforms of the Green's dyadics for a three-dimensional problem:

$$\overline{\mathbf{G}}^E = F_y\{\tilde{\mathbf{G}}^E\}$$

$$\overline{\mathbf{G}}^M = F_y\{\tilde{\mathbf{G}}^M\}$$

150   *Electromagnetic methods*

Accordingly, the dyadic components operating in equations (6.6.25)–(6.6.28) are also Fourier transforms of the dyadic components expressed by equations (6.6.7) and (6.6.8).

To solve the unknown electric and magnetic source vector functions $\bar{\mathbf{a}}$ and $\bar{\mathbf{b}}$ on line s, the boundary conditions to $\bar{\mathbf{E}}$ and $\bar{\mathbf{H}}$ are enforced by substituting equations (6.6.25)–(6.6.28) into equations (6.6.21)–(6.6.24). Treating the effect of discontinuity in the integration of the $\nabla \times$ terms in equations (6.6.25)–(6.6.28) as the calculation point $\mathbf{r}$ is moved to the source contour s, and assuming quasi-static conditions, $g \gg \omega\varepsilon$, which are generally valid in geophysical applications, the following sets of line integral equations are obtained for the y and t components of the transformed source densities $\bar{\mathbf{a}}$ and $\bar{\mathbf{b}}$:

$$\frac{-i\omega(\mu_1 + \mu_2)\bar{b}_t(\mathbf{r})}{2}$$

$$-\mathbf{y} \cdot \left[\int_s \left\{ \left[\overline{G}_{02}^{ee}(\mathbf{r},\mathbf{r}_0) - \overline{G}_{01}^{ee}(\mathbf{r},\mathbf{r}_0) + \overline{G}_S^{ee}(\mathbf{r},\mathbf{r}_0) \right] \cdot \mathbf{y}\bar{a}_y(\mathbf{r}_0) \right. \right.$$

$$+ \left[\overline{G}_{02}^{ee}(\mathbf{r},\mathbf{r}_0) - \overline{G}_{01}^{ee}(\mathbf{r},\mathbf{r}_0) + \overline{G}_S^{ee}(\mathbf{r},\mathbf{r}_0) \right] \cdot \mathbf{t}\bar{a}_t(\mathbf{r}_0)$$

$$+ \left[\overline{G}_{02}^{em}(\mathbf{r},\mathbf{r}_0) - \overline{G}_{01}^{em}(\mathbf{r},\mathbf{r}_0) + \overline{G}_S^{em}(\mathbf{r},\mathbf{r}_0) \right] \cdot \mathbf{y}\bar{b}_y(\mathbf{r}_0)$$

$$\left. \left. + \left[\overline{G}_{02}^{em}(\mathbf{r},\mathbf{r}_0) - \overline{G}_{01}^{em}(\mathbf{r},\mathbf{r}_0) + \overline{G}_S^{em}(\mathbf{r},\mathbf{r}_0) \right] \cdot \mathbf{t}\bar{b}_t(\mathbf{r}_0) \right\} ds_0 \right]$$

$$= \bar{E}_{Py}(\mathbf{r}), \qquad \mathbf{r} \text{ on } s \tag{6.6.29}$$

$$\frac{-i\omega(\mu_1 + \mu_2)\bar{b}_y(\mathbf{r})}{2}$$

$$+\mathbf{t} \cdot \left[\int_s \left\{ \left[\overline{G}_{02}^{ee}(\mathbf{r},\mathbf{r}_0) - \overline{G}_{01}^{ee}(\mathbf{r},\mathbf{r}_0) + \overline{G}_S^{ee}(\mathbf{r},\mathbf{r}_0) \right] \cdot \mathbf{y}\bar{a}_y(\mathbf{r}_0) \right. \right.$$

$$+ \left[\overline{G}_{02}^{ee}(\mathbf{r},\mathbf{r}_0) - \overline{G}_{01}^{ee}(\mathbf{r},\mathbf{r}_0) + \overline{G}_S^{ee}(\mathbf{r},\mathbf{r}_0) \right] \cdot \mathbf{t}\bar{a}_t(\mathbf{r}_0)$$

$$+ \left[\overline{G}_{02}^{em}(\mathbf{r},\mathbf{r}_0) - \overline{G}_{01}^{em}(\mathbf{r},\mathbf{r}_0) + \overline{G}_S^{em}(\mathbf{r},\mathbf{r}_0) \right] \cdot \mathbf{y}\bar{b}_y(\mathbf{r}_0)$$

$$\left. \left. + \left[\overline{G}_{02}^{em}(\mathbf{r},\mathbf{r}_0) - \overline{G}_{01}^{em}(\mathbf{r},\mathbf{r}_0) + \overline{G}_S^{em}(\mathbf{r},\mathbf{r}_0) \right] \cdot \mathbf{t}\bar{b}_t(\mathbf{r}_0) \right\} ds_0 \right]$$

$$= -\bar{E}_P(\mathbf{r}) \cdot \mathbf{t}, \qquad \mathbf{r} \text{ on } s \tag{6.6.30}$$

$$\frac{(g_1 + g_2)\,\bar{a}_t(\mathbf{r})}{2} - \mathbf{y} \cdot \Bigg[\int_s \Big\{ \Big[\overline{\mathbf{G}}_{02}^{me}(\mathbf{r}, \mathbf{r}_0) - \overline{\mathbf{G}}_{01}^{me}(\mathbf{r}, \mathbf{r}_0) + \overline{\mathbf{G}}_{S}^{me}(\mathbf{r}, \mathbf{r}_0) \Big] \cdot \mathbf{y}\bar{a}_y(\mathbf{r}_0)$$

$$+ \Big[\overline{\mathbf{G}}_{02}^{me}(\mathbf{r}, \mathbf{r}_0) - \overline{\mathbf{G}}_{01}^{me}(\mathbf{r}, \mathbf{r}_0) + \overline{\mathbf{G}}_{S}^{me}(\mathbf{r}, \mathbf{r}_0) \Big] \cdot \mathbf{t}\bar{a}_t(\mathbf{r}_0)$$

$$+ \Big[\overline{\mathbf{G}}_{02}^{mm}(\mathbf{r}, \mathbf{r}_0) - \overline{\mathbf{G}}_{01}^{mm}(\mathbf{r}, \mathbf{r}_0) + \overline{\mathbf{G}}_{S}^{mm}(\mathbf{r}, \mathbf{r}_0) \Big] \cdot \mathbf{y}\bar{b}_y(\mathbf{r}_0)$$

$$+ \Big[\overline{\mathbf{G}}_{02}^{mm}(\mathbf{r}, \mathbf{r}_0) - \overline{\mathbf{G}}_{01}^{mm}(\mathbf{r}, \mathbf{r}_0) + \overline{\mathbf{G}}_{S}^{mm}(\mathbf{r}, \mathbf{r}_0) \Big] \cdot \mathbf{t}\bar{b}_t(\mathbf{r}_0) \Big\} \, ds_0 \Bigg]$$

$$= \overline{H}_{Py}(\mathbf{r}), \quad \mathbf{r} \text{ on } s \tag{6.31}$$

$$\frac{(g_1 + g_2)\,\bar{a}_y(\mathbf{r})}{2}$$

$$+ \mathbf{t} \cdot \Bigg[\int_s \Big\{ \Big[\overline{\mathbf{G}}_{02}^{me}(\mathbf{r}, \mathbf{r}_0) - \overline{\mathbf{G}}_{01}^{me}(\mathbf{r}, \mathbf{r}_0) + \overline{\mathbf{G}}_{S}^{me}(\mathbf{r}, \mathbf{r}_0) \Big] \cdot \mathbf{y}\bar{a}_y(\mathbf{r}_0)$$

$$+ \Big[\overline{\mathbf{G}}_{02}^{me}(\mathbf{r}, \mathbf{r}_0) - \overline{\mathbf{G}}_{01}^{me}(\mathbf{r}, \mathbf{r}_0) + \overline{\mathbf{G}}_{S}^{me}(\mathbf{r}, \mathbf{r}_0) \Big] \cdot \mathbf{t}\bar{a}_t(\mathbf{r}_0)$$

$$+ \Big[\overline{\mathbf{G}}_{02}^{mm}(\mathbf{r}, \mathbf{r}_0) - \overline{\mathbf{G}}_{01}^{mm}(\mathbf{r}, \mathbf{r}_0) + \overline{\mathbf{G}}_{S}^{mm}(\mathbf{r}, \mathbf{r}_0) \Big] \cdot \mathbf{y}\bar{b}_y(\mathbf{r}_0)$$

$$+ \Big[\overline{\mathbf{G}}_{02}^{mm}(\mathbf{r}, \mathbf{r}_0) - \overline{\mathbf{G}}_{01}^{mm}(\mathbf{r}, \mathbf{r}_0) + \overline{\mathbf{G}}_{S}^{mm}(\mathbf{r}, \mathbf{r}_0) \Big] \cdot \mathbf{t}\bar{b}_t(\mathbf{r}_0) \Big\} \, ds_0 \Bigg]$$

$$= -\overline{\mathbf{H}}_P(\mathbf{r}) \cdot \mathbf{t}, \quad \mathbf{r} \text{ on } s \tag{6.6.32}$$

The components $\bar{a}_y$, $\bar{a}_t$ and $\bar{b}_y$, $\bar{b}_t$ of the source densities are obtained numerically from equations (6.6.29)–(6.6.32) for a sufficient set of k_y values. The transform-domain fields $\overline{\mathbf{E}}$ and $\overline{\mathbf{H}}$ corresponding to the chosen set of k_y values are then calculated at the required points $\mathbf{r}$ of the plane $y = 0$ by using equations (6.6.25)–(6.6.28). The space-domain fields $\mathbf{E}$ and $\mathbf{H}$ can then be calculated from this set at any arbitrary point $\mathbf{R}$ by the inverse Fourier transformation (6.6.20b). In the numerical computations carried out in [44], it was found that field transforms at each field point are required at between 15 and 21 values of k_y, before inverse Fourier transformation can be performed, yielding the space-domain field values.

Let us consider, after [44], the magnetic field anomalies generated by a vertical coaxial loop system over an anomalous body buried in a homogeneous half-space. The results are calculated at frequencies 32 Hz and 263 Hz by using the $2\frac{1}{2}$-dimensional surface integral equations stated above. The set of surface integral equations is solved numerically by means of the point-matching method on the subsectional basis. The electric and magnetic source densities are taken to vary linearly along the subsections, 78 in number. The specifications of the model are given in more detail in Figure 6.6.

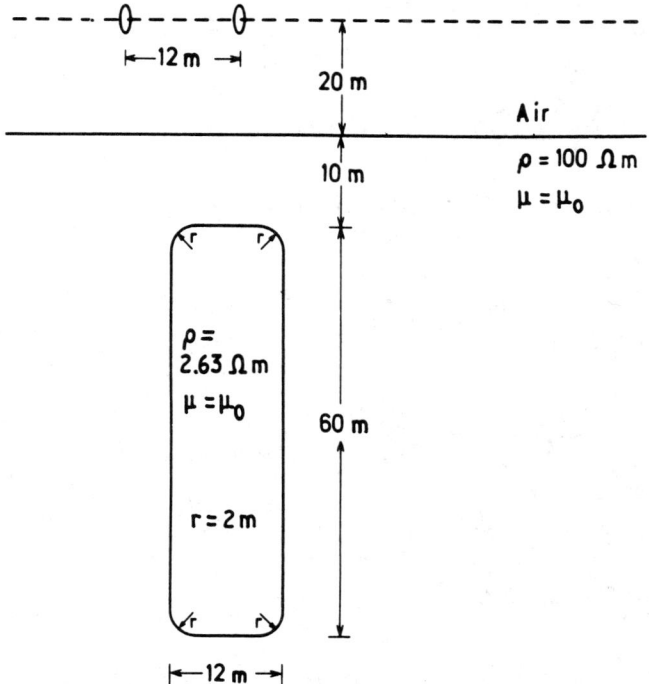

Figure 6.6 A vertical coaxial loop profile over a two-dimensional conductor buried in a half-space [44, p. 660].

Figure 6.7 shows the integral equation modelling results together with the results obtained by finite element modelling. The surface integral equation results and the finite element results agree very well at 263 Hz. At 32 Hz the imaginary anomalies also agree well. However, the real anomaly obtained by integral equation modelling at the lower frequency of 32 Hz is considerably in error; this is attributed by Doherty to matrix ill-conditioning resulting from a non-unique solution at zero frequency for the real response due to magnetic sources.

6.7 INTEGRAL EQUATION SOLUTION FOR ELECTROMAGNETIC FIELDS IN A THIN CONDUCTOR MODEL

In this section we describe the solution of an electromagnetic integral equation for a thin conductor model. This is done by substituting two potential functions for the anomalous current in the conductor, in order to separate the current into its solenoidal and laminar components. The idea behind this technique was presented by Lajoie and West [65]. Published computer programs for

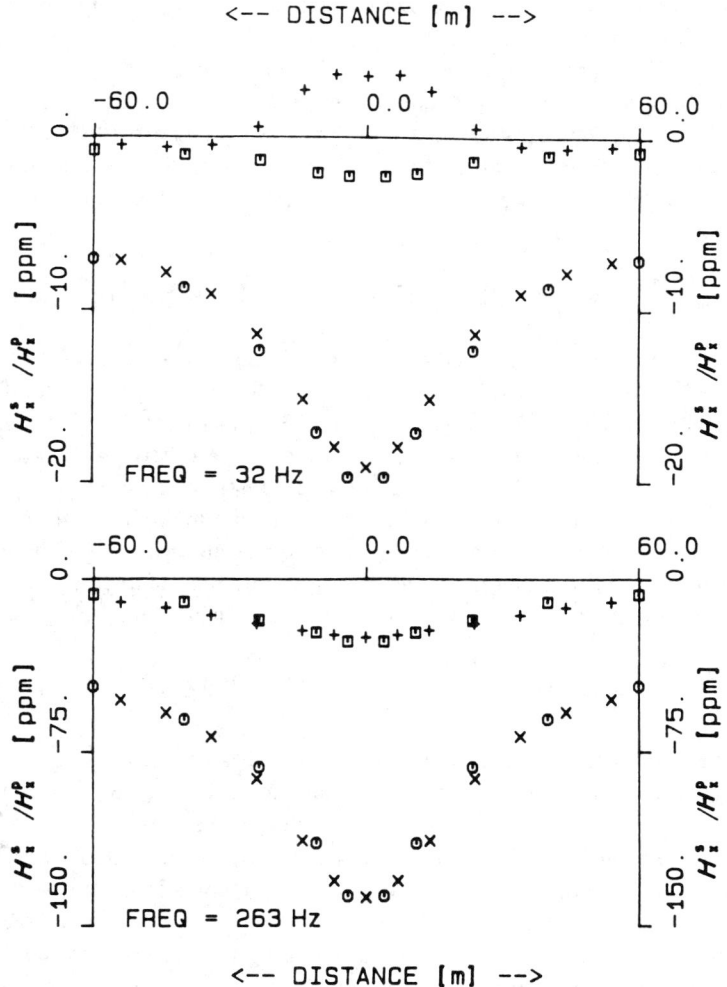

Figure 6.7 Computed results for model of Fig. 6.6. Finite element results shown as real component (□) and imaginary component (○); surface integral equation results shown as real component (+) and imaginary component (×) [44, p. 661].

electromagnetic modelling of a thin conducting plate located in free space are also reported by Dyck *et al.* [66]. A thin conductor model is useful in simulating such geological features as mineralized veins and fracture zones, in which one dimension of the anomalous body is generally much shorter than the other two. This makes it possible to limit the number of the unknown field components to two, both aligned in the surface containing the conductor.

Corresponding solutions for the static electric and magnetic problems were given in Sections 3.3.6 and 5.2.3.

The specifications of the model are illustrated in Fig. 3.5(a), where the system of coordinates k, l, n is fixed to surface A, describing the thin conductor, such that k- and l-axes are parallel to and the n axis normal to A. We begin the formulation from the perfectly three-dimensional volume integral representation for the electric field, given in accordance with equation (6.4.7) as follows:

$$\mathbf{E}(\mathbf{R}) = \mathbf{E}_P(\mathbf{R}) + \int_V \tilde{\mathbf{G}}(\mathbf{R}, \mathbf{R}_0) \cdot \Delta k^2(\mathbf{R}_0) \, \mathbf{E}(\mathbf{R}_0) \, dV_0 \qquad (6.7.1)$$

where V is the volume of the conductor, and $\Delta k^2 = k_1^2 - k_2^2$ with $k_j^2 = -i\omega\mu_j(g_j + i\omega\varepsilon_j)$, $g_j = 1/\rho_j$, $j = 1, 2$. $\tilde{\mathbf{G}}$ is the electric Green's dyadic, specified by equations (6.3.2)–(6.3.6) and (6.3.11)–(6.3.13). Let us assume that the thickness t of the conductor is very small relative to the other dimensions of the model. Then the n-component of $\mathbf{E}$ perpendicular to the conductor has no effect on the secondary field $\mathbf{E}_S$ and can be neglected in the integral. Assuming that t is also small in relation to the depth of penetration in the conductor, and performing the integration over the thickness t of the conductor, the volume integral in equation (6.7.1) is transformed into a surface integral as follows:

$$\mathbf{E}(\mathbf{R}) = \mathbf{E}_P(\mathbf{R}) - i\omega\mu \int_A \tilde{\mathbf{G}}(\mathbf{R}, \mathbf{R}_0) \cdot S(\mathbf{R}_0) \, \mathbf{E}_A(\mathbf{R}_0) \, dA_0 \qquad (6.7.2)$$

where A is the area of the conductor, $\mathbf{E}_A = \mathbf{n} \times \mathbf{E}$ is the two-dimensional projection of electric field $\mathbf{E}$ on conducting surface A (see Fig. 3.5), and $S = t\Delta(g + i\omega\varepsilon)$ is the longitudinal conductance of the conductor.

When equation (6.7.2) is used to calculate the field $\mathbf{E}$ outside A, the Green's dyadic $\tilde{\mathbf{G}}$ of the equation involves nine components, as given in equation (6.3.2). However, if conductor A is tabular, the equation reduced from equation (6.7.2) for obtaining the two unknown longitudinal components of $\mathbf{E}$ in A is a two-dimensional integral equation with a (2×2) Green's dyadic:

$$\mathbf{E}_A(\mathbf{r}) = \mathbf{E}_{AP}(\mathbf{r}) - i\omega\mu \int_A \mathbf{G}_A(\mathbf{r}, \mathbf{r}_0) \cdot S(\mathbf{r}_0) \, \mathbf{E}_A(\mathbf{r}_0) \, dA_0 \qquad (6.7.3)$$

where both $\mathbf{r}$ and $\mathbf{r}_0$ are in A. Green's dyadic of equation (6.7.3) operating in plane A can be written in the system of coordinates a, b, n in the following form:

$$\mathbf{G}_A = g_{aa} \, \mathbf{aa} + g_{ab} \, \mathbf{ba}$$
$$+ g_{ba} \, \mathbf{ab} + g_{bb} \, \mathbf{bb} \qquad (6.7.4)$$

It was found in the numerical computations done in [65] that the components of $\mathbf{E}_A$ obtained from equation (6.7.3) seemed reasonable in form, but that their amplitude and phase responses were quite inappropriate for any reasonably resistive host medium. This was explained as a result of two main causes. First, when the conductivity g_2 of the host medium is low, the $1/k_2^2$ multiplier in the laminar part of Green's dyadic described by equation (6.3.12) is so large that the contribution of the solenoidal current component is effectively lost when the two terms are added together in the matrix equation obtained from equation (6.7.3) by discretization. Second, numerical problems are a result of mixing the solenoidal and the laminar components in each spatial component of field $\mathbf{E}_A$.

In order to increase the stability of the numerical solution, field $\mathbf{E}_A$ and thus the anomalous current in A are partitioned into and expressed in terms of solenoidal (divergence-free) and laminar (curl-free) components [65]. Since it is known from vector analysis that any vector field may be expressed as the sum of the gradient of a scalar potential and the curl of a vector potential, the following substitution may be made in equation (6.7.3):

$$\mathbf{E}_A = (\nabla \times \mathbf{n}U)_A - \nabla_A V \qquad (6.7.5)$$

where $\mathbf{n}$ is the normal unit vector of surface A, and ∇_A denotes the two-dimensional differential operator in plane A. The vector potential $\mathbf{n}U$ represents the solenoidal currents due to induction and any current vortex associated with it is constrained to flow within A. Therefore, the boundary condition $U = 0$ must be satisfied on the edges s of A. The scalar potential V represents the laminar conduction currents entering and leaving conductor A. Potential V approaches a constant value (zero) as the resistivity of the host material increases, reducing the numerical problems caused by the multiplier $1/k_2^2$ involved in Green's dyadic. The direction of the vector potential is fixed and the problem thus takes the form of solving for two scalar functions, U and V, in A.

A set of integral equations for obtaining U and V in A can be obtained, for example, by substituting equation (6.7.5) into equation (6.7.3), taking separately the curl and the divergence of the equation thus obtained, and taking into account that $\nabla_A \cdot (\nabla \times \mathbf{n}U)_A = \nabla \times (\nabla_A U) = 0$ in A and $U = 0$ on edges s of A.

After U and V have been solved in A, the electric field $\mathbf{E}$ can be calculated at any arbitrary point of the model by using equation (6.7.2) with substitution of potentials U and V according to equation (6.7.5).

The magnetic field in the model is obtained as follows:

$$\mathbf{H}(\mathbf{R}) = \mathbf{H}_P(\mathbf{R}) + \int_A \nabla \times \widetilde{\mathbf{G}}(\mathbf{R}, \mathbf{R}_0) \cdot S(\mathbf{R}_0) \, \mathbf{E}_A(\mathbf{R}_0) \, dA_0 \qquad (6.7.6)$$

In the limit as ω approaches zero, equation (6.7.2) reduces to equation (3.3.60), representing the potential of a static electric field.

Numerical results were computed in [65] by using the point-matching method. Appropriate spline functions were used to represent the potential values between the grid points. The specific forms of the discretized integral equations for solving U and V and a closer consideration of the numerical solution of these equations can be found in [65].

156 *Electromagnetic methods*

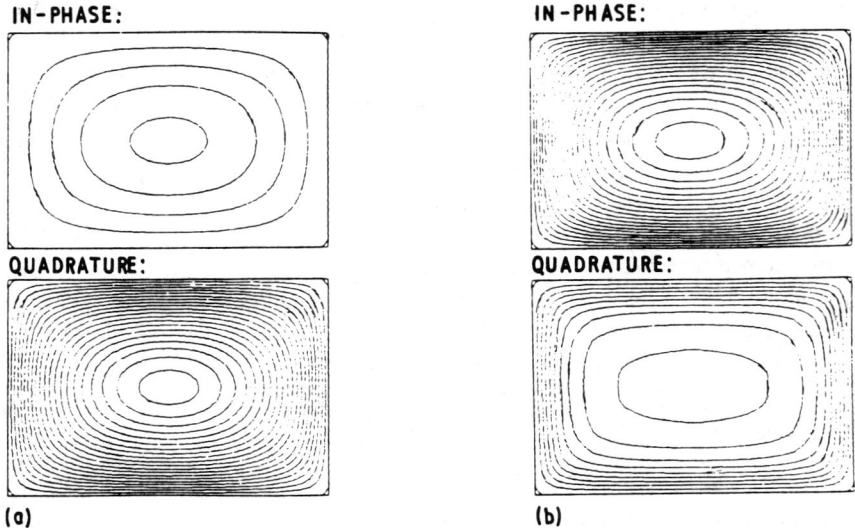

Figure 6.8 Contours of SU for a plate having dimensions 225 m × 325 m and located in free space in a uniform magnetic primary field perpendicular to the plate. (a) $\omega S\mu L = 2.44$, contour interval 1.531 A, and (b) $\omega S\mu L = 24.4$, contour interval 6.267 A. After [65].

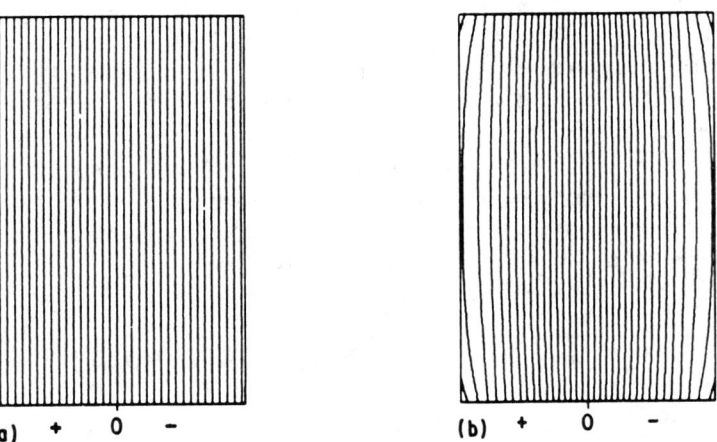

Figure 6.9 Contours of SV for a plate in a uniform space having conductivity 0.001 S m^{-1} and excited by a homogeneous electric field parallel to the width of the plate. (a) $S = 0.001$ S, contour interval 0.0057 A; and (b) $S = 10^{10}$ S, contour interval 1 A. After [65].

Integral equation solution 157

Figures 6.8 and 6.9 are plotted to illustrate the general properties of potentials U and V. Figure 6.8 shows contours of the real and imaginary components of a product SU in a rectangular plate located in a free space. Current is induced by an oscillating uniform magnetic field perpendicular to the plate. The results are given for two different induction numbers $\omega S\mu L$, where L is the long dimension of the plate, the phase references being taken relative to the primary magnetic field. The contours of U may be visualized as lines of current flow within the plate. At the lower induction number (Fig. 6.8(a)), the current is mainly in the imaginary component, whereas at the higher induction number (Fig. 6.8(b)), the real component of current dominates. With a further increase in the induction number the current flow in the plate becomes saturated (not shown), making the current entirely real and more concentrated near the edges of the plate.

Figure 6.9 shows contours of a product SV at two different S values in a rectangular plate embedded in a homogeneous medium of conductivity $0.001\,S\,m^{-1}$. The primary field is a static homogeneous electric field parallel to the shorter dimension of the plate. The laminar surface current density in the plate can be written as $t\mathbf{j} = S\nabla_A V$. The current in the plate can thus be visualized as flowing perpendicular to the contours. At the lower S value (Fig. 6.9(a)) the SV contours are straight equidistant lines, mainly representing the homogeneous flow of the primary current. At the higher value of S (Fig. 6.9(b)), the potential is distorted due to the effects of current collection in the plate between the centre and the edge located at the higher primary potential, and of current discharge between the centre and the other edge. As S tends to infinity (not shown), the plate becomes electrostatically saturated so that V approaches zero, while the product SV and current density $S\nabla_A V$ approach finite limiting values within the plate.

7
Seismic methods

7.1 INTRODUCTION

Seismic methods make use of the fact that elastic waves propagate with different velocities in different rocks (see [9, Table 7.2]). A seismic survey is conducted by generating a seismic disturbance in the earth and then recording the transmitted scattered waves with a series of geophones. The disturbance propagates from the source point into the surrounding medium as two isolated wave modes, namely **compressional** (P) and **shear** (S) waves that expand with characteristic velocities c_P and c_S, respectively. In a homogeneous isotropic medium, such a P or S wave continues to propagate as a P or S wave, but should discontinuity surfaces be present, the waves are scattered (refracted, reflected or diffracted). Scattering processes are also associated with mode conversion, resulting in the conversion of P waves into S waves and vice versa. The times required by scattering waves to travel from discontinuities to the geophones are used to derive subsurface structural information. Further description of the techniques of measurement, processing and interpretation in seismic surveys may be found in [9] and [33].

The most extensive applications of seismic methods are in oil exploration, where they are used primarily for indirect mapping of geological structures and stratigraphic units. Other applications include engineering surveys and the investigation of water resources. Seismic techniques are not readily applicable to mineral exploration because they do not yield high definition where boundaries between different rock units are highly irregular. However, they can be employed in locating geological features such as buried channels delineated by heavy mineral accumulation.

Integral equation techniques are applied in both forward and backward modelling of seismic data [67]. In forward modelling, the wave field is constructed at some future time for a specific location. The model results are then compared with actual seismic recording results in order to test hypotheses concerning the subsurface being surveyed. In inverse modelling, generally referred to as **migration**, the subsurface wave field is reconstructed for previous times. In this case the time that is usually of interest is the instant at which the registered wave reached a particular scattering horizon, the final purpose of migration being to produce an image of the subsurface from the recordings obtained at the surface.

The integral representation of Huygens's principle for acoustic waves satisfying the scalar wave equation was first given for the frequency domain by Helmholtz in 1860, and for the time domain by Kirchhoff in 1883. The frequency-dependent functions of Helmholtz integrals are related to the corresponding time-dependent functions of Kirchhoff integrals by the Fourier transformation. Full elastic wave fields are, however, both physically and formally more difficult since they are composed of both a compressional (P) and a rotational (S) mode that propagate at different velocities. Hence the elastic displacement vector does not satisfy the conventional wave equation.

In this chapter we shall deal with the methods by which the generalized forms of Hooke's law and Newton's law can be converted into Helmholtz and Kirchhoff integral formulas. Considerations of the full elastic wave fields are essentially based on the work of Pao and Varatharajulu [68]. For background to the elements of linear strain theory and application of the Helmholtz and Kirchhoff integral formulas in forward and inverse modelling of stress fields in seismic surveys, reference is made to textbooks more closely concerned with application of the wave theory in seismic exploration, such as Berkhout [69] and Morgan [67].

7.2 INTEGRAL FORMULAS FOR ELASTIC WAVE FIELDS IN AN ANISOTROPIC MEDIUM

Consider an anisotropic earth model that is composed of two homogeneous regions, V_1 and V_2, separated by surface S_1 and having different densities ρ and elastic properties Γ (see Fig. 7.1). The elastic properties are defined by the fourth-rank elastic stiffness tensor, denoted in the Cartesian system of

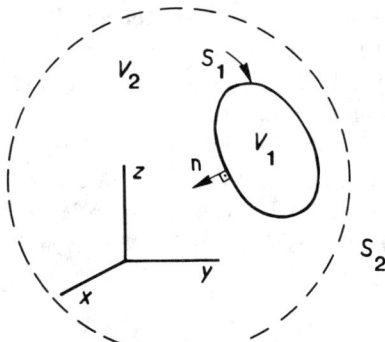

Figure 7.1 Elastic earth model composed of two regions, V_1 and V_2, with different densities and elastic constants.

160 *Seismic methods*

coordinates x, y, z by Γ_{ijkl}, where $i, j, k, l = 1, 2, 3$ stand for x, y, z, respectively. The stiffness tensor possesses the following symmetry:

$$\Gamma_{ijkl} = \Gamma_{jikl} = \Gamma_{ijlk} = \Gamma_{klij} \tag{7.2.1}$$

The constitutive relation for the medium is given by the generalized Hooke's law [68]:

$$T_{ij} - \Gamma_{ijkl}(\partial_k u_l) = 0 \tag{7.2.2}$$

where u_l represents the three components of the displacement vector $\mathbf{u}$, and T_{ij} the nine components of the symmetric stress tensor $\mathbf{T}$. The notation assumes that in the products involving repeated subscripts, summation is performed over 1, 2 and 3 relative to the repeated subscripts.

The equation of motion for harmonic waves having time dependence $\exp(i\omega t)$ is given by the generalized Newton's law as follows:

$$\partial_i T_{ij} + \rho \omega^2 u_j = -\rho f_j \tag{7.2.3}$$

where ρ is density and $\mathbf{f}$ is the body force per unit mass. Eliminating $\mathbf{T}$ from equations (7.2.2) and (7.2.3), we obtain the equation of motion for displacement $\mathbf{u}$:

$$\partial_i(\Gamma_{ijkl}\partial_k u_l) + \rho \omega^2 u_j = -\rho f_j \tag{7.2.4}$$

The equations corresponding to equation (7.2.2)–(7.2.4) for Green's tensors are as follows:

$$\Sigma_{ijm} - \Gamma_{ijkl}(\partial_k G_{lm}) = 0 \tag{7.2.5}$$

$$\partial_i \Sigma_{ijm} + \rho \omega^2 G_{jm} = -\delta_{jm}\delta(\mathbf{R} - \mathbf{R}_0) \tag{7.2.6}$$

$$\partial_i(\Gamma_{ijkl}\partial_k G_{lm}) + \rho \omega^2 G_{jm} = -\delta_{jm}\delta(\mathbf{R} - \mathbf{R}_0) \tag{7.2.7}$$

where δ_{jm} is the Kronecker delta, and $\delta(\mathbf{R} - \mathbf{R}_0)$ is the Dirac delta function. The second-rank tensor $\mathbf{G}$ is referred to as Green's **displacement tensor**, and the third-rank tensor Σ as Green's **stress tensor**. Physically, $\mathbf{G}(\mathbf{R}, \mathbf{R}_0)$ represents the displacement field at $\mathbf{R}$ due to three mutually perpendicular concentrated forces applied at $\mathbf{R}_0$. Expressed in component form, $G_{mn}(\mathbf{R}, \mathbf{R}_0)$ represents the displacement component at $\mathbf{R}$ in the direction of the x_n-axis due to a force applied at $\mathbf{R}_0$ in the direction of the x_m-axis. Likewise, tensor $\Sigma(\mathbf{R}, \mathbf{R}_0)$ represents the stress field at $\mathbf{R}$ generated by three mutually perpendicular concentrated forces at $\mathbf{R}_0$. Hence, the role of Σ with respect to $\mathbf{T}$ is analogous to that of $\mathbf{G}$ with respect to $\mathbf{u}$. Tensors $\mathbf{G}$ and Σ satisfy the following symmetry relations:

$$G_{im}(\mathbf{R}, \mathbf{R}_0) = G_{mi}(\mathbf{R}_0, \mathbf{R}) \tag{7.2.8}$$

$$\Sigma_{ijm}(\mathbf{R}, \mathbf{R}_0) = \Sigma_{jim}(\mathbf{R}_0, \mathbf{R}) \tag{7.2.9}$$

Integral formulas for elastic wave fields in an anisotropic medium 161

where subscript m ($m = 1, 2, 3$) refers to the direction of the body force at $\mathbf{R}_0$.

To transform the wave field represented by equations (7.2.2)–(7.2.9) into an integral representation we form the vector function

$$\mathbf{Q} = \mathbf{T} \cdot \mathbf{G} - \mathbf{u} \cdot \Sigma \qquad (7.2.10)$$

We then substitute $\mathbf{Q}$ into Gauss's theorem

$$\int_V \nabla \cdot \mathbf{Q} \, dV = \oint_S \mathbf{Q} \cdot \mathbf{n} \, dS \qquad (7.2.11)$$

and apply the theorem to region V_1 and its boundary S_1 (see Fig. 7.1). Substituting equations (7.2.2), (7.2.3), (7.2.5) and (7.2.6), and applying the symmetry relations from equation (7.2.1), we obtain the full elastic Helmholtz integral formula:

$$\int_{S_1} n_{0j} \Gamma_{ijkl} [G_{im}(\mathbf{R}, \mathbf{R}_0) \partial_{0k} u_l(\mathbf{R}_0) - u_i(\mathbf{R}_0) \partial_{0k} G_{lm}(\mathbf{R}, \mathbf{R}_0)] \, dS_0$$

$$+ \int_{V_1} \rho f_i(\mathbf{R}_0) G_{im}(\mathbf{R}, \mathbf{R}_0) \, dV_0 = \begin{cases} u_m(\mathbf{R}), & \mathbf{R} \in V_1 \\ 0, & \mathbf{R} \in V_2 \end{cases} \qquad (7.2.12)$$

The elastic constants and density involved in this equation are those of region V_1.

The corresponding integral representation for the stress field in region V_2 can be obtained by applying equation (7.2.11) to region V_2 and its internal and external surfaces S_1 and S_2 (again see Fig. 7.1), and letting S_2 recede to infinity. Imposing the radiation condition (given more specifically in Section 7.3), which makes the surface integral over S_2 vanish as S_2 goes to infinity, and assuming that all body forces are concentrated into region V_1, the elastic Helmholtz integral formula for the region external to S_1 takes the following form:

$$\int_{S_1} n_{0j} \Gamma_{ijkl} [u_i(\mathbf{R}_0) \partial_{0k} G_{lm}(\mathbf{R}, \mathbf{R}_0) - G_{im}(\mathbf{R}, \mathbf{R}_0) \partial_{0k} u_l(\mathbf{R}_0)] \, dS_0$$

$$= \begin{cases} u_m(\mathbf{R}), & \mathbf{R} \in V_2 \\ 0, & \mathbf{R} \in V_1 \end{cases} \qquad (7.2.13)$$

The density and the elastic constants in equation (7.2.13) are those of region V_2. If $\mathbf{u}(\mathbf{R}_0)$ is considered as the secondary source on the wave surface on S_1, equation (7.2.13) is the mathematical statement of Huygens's principle for elastic waves in the frequency domain.

Consider now that elastic scattering takes place at boundary S_1 (Fig. 7.1). The earth materials in regions V_1 and V_2 are characterized by densities ρ_1 and ρ_2, and stiffness tensors Γ^1 and Γ^2, respectively. The sources of the incident wave $\mathbf{u}^i(\mathbf{R})$ are assumed to be located in V_2. As a result of the discontinuity in ρ and Γ, the incident wave is scattered at the surface S_1. The total wave field outside S_1 can be expressed in terms of the incident and the scattered waves

162 *Seismic methods*

$\mathbf{u}^i$ and $\mathbf{u}^s$ as follows:

$$\mathbf{u}(\mathbf{R}) = \mathbf{u}^i(\mathbf{R}) + \mathbf{u}^s(\mathbf{R}) \qquad (7.2.14)$$

when $\mathbf{R}$ is in V_2. Inside S_1 there is a standing wave, hence $\mathbf{u}(\mathbf{R}) = \mathbf{u}^i(\mathbf{R})$ in V_1. Applying equation (7.2.12) to the incident field outside S_1 yields:

$$\int_{S_1} n_{0j} \Gamma^2_{ijkl} [G^2_{im}(\mathbf{R}, \mathbf{R}_0) \partial_{0k} u^i_l(\mathbf{R}_0)$$

$$- u^i_i(\mathbf{R}_0) \partial_{0k} G^2_{lm}(\mathbf{R}, \mathbf{R}_0)] \mathrm{d}S_0 = 0, \quad \mathbf{R} \in V_2 \qquad (7.2.15)$$

where G^2 satisfies equation (7.2.7) with density and elastic constants ρ_2 and Γ^2.

The scattered field $\mathbf{u}^s$ is emitted by secondary sources on or inside S_1. Applying equation (7.2.13) to the scattered field outside S_1, we obtain:

$$\int_{S_1} n_{0j} \Gamma^2_{ijkl} [u^s_i(\mathbf{R}_0) \partial_{0k} G^2_{lm}(\mathbf{R}, \mathbf{R}_0) - G^2_{im}(\mathbf{R}, \mathbf{R}_0) \partial_{0k} u^s_l(\mathbf{R}_0)] \mathrm{d}S_0$$

$$= u^s_m(\mathbf{R}), \quad \mathbf{R} \in V_2 \qquad (7.2.16)$$

Combining equations (7.2.15) and (7.2.16) and substituting equation (7.2.14), we obtain for the total wave field outside surface S_1:

$$\int_{S_1} n_{0j} \Gamma^2_{ijkl} [u_i(\mathbf{R}_0) \partial_{0k} G^2_{lm}(\mathbf{R}, \mathbf{R}_0) - G^2_{im}(\mathbf{R}, \mathbf{R}_0) \partial_{0k} u_l(\mathbf{R}_0)] \mathrm{d}S_0$$

$$+ u^i_m(\mathbf{R}) = u_m(\mathbf{R}), \quad \mathbf{R} \in V_2 \qquad (7.2.17)$$

The field inside the surface S_1 can be calculated from equation (7.2.12):

$$\int_{S_1} n_{0j} \Gamma^1_{ijkl} [G^1_{im}(\mathbf{R}, \mathbf{R}_0) \partial_{0k} u_l(\mathbf{R}_0) - u_i(\mathbf{R}_0) \partial_{0k} G^1_{lm}(\mathbf{R}, \mathbf{R}_0)] \mathrm{d}S_0$$

$$= u^1_m(\mathbf{R}), \quad \mathbf{R} \in V_1 \qquad (7.2.18)$$

The $\mathbf{G}^1$ is Green's displacement tensor with material constants ρ_1 and Γ^1, with $\mathbf{u}(\mathbf{R}_0)$ and its derivative being the same as in equation (7.2.17).

If $\mathbf{u}$ and its derivative are known on S_1, the total field can be calculated outside and inside S_1 using equations (7.2.17) and (7.2.18), respectively. For most problems, however, either or both of them are unknown. If one of them is known, the other can be derived from an integral equation obtained by moving the point $\mathbf{R}$ in equation (7.2.17) to surface S_1. If neither surface source is known, a system of integral equations for $\mathbf{u}$ and its derivative can be formulated from equations (7.2.17) and (7.2.18) by the moving point $\mathbf{R}$ to surface S_1.

Equations (7.2.12)–(7.2.18) can be applied only to classes of anisotropy for which Green's displacement tensor $\mathbf{G}$ is known. $\mathbf{G}$ can be obtained as the solution of wave equation (7.2.7). However, this solution is very complicated, because in a completely anisotropic medium, Γ is a 6×6 symmetric tensor having 21 independent components. Thus for a given direction of wave propagation there are three phase velocities and three corresponding displacement vectors, and the displacement vectors are neither absolutely compressional nor rotational in nature.

Elastic anisotropic Kirchhoff–Helmholtz integral formulas have so far had only limited application in seismic exploration. However, as the requirements for seismic surveys in oil and gas exploration become more and more stringent in the future, increased use of the elastic wave equation as the basis of seismic modelling is to be expected. This trend has already become apparent for forward modelling, especially in modelling the true vertical field component at the earth's surface. However, since full application of this theory in inverse modelling requires three-component recording of data, resulting in a considerable increase in expense, its usage may remain more limited. As an example of the application of the full elastic Helmholtz integral formula, the recent work of Wapenaar and Haimé [70] should be mentioned, where this integral was used as the basis for deriving amplitude extrapolation operators for primary P and S waves.

Since the possibilities of obtaining solutions for Green's tensors of anisotropic elastic integral formulas are very restricted, in Section 7.3 we shall consider a considerably simpler earth model, namely an isotropic elastic model. As will be seen, the procedures by which the integral formulas for an isotropic model are obtained from the generalized Hooke's and Newton's laws are completely analogous to those used in deriving the integral formulas for an anisotropic model.

7.3 INTEGRAL FORMULAS FOR ELASTIC WAVE FIELDS IN AN ISOTROPIC MEDIUM

Assume that the elastic properties of media 1 and 2 in an earth model, illustrated in Fig. 7.1, are isotropic. Whereas 21 independent constants were required to represent the elastic properties of an anisotropic material, the elastic properties of an isotropic material can be described with the aid of only two independent constants, known as **Lamé's constants** λ and μ.

For isotropic solids, the elastic stiffness tensor Γ can be expressed in the following form [68]:

$$\Gamma_{ijkl} = \lambda \, \delta_{ij}\delta_{kl} + \mu(\delta_{ik}\delta_{jl} + \delta_{il}\delta_{kj}) \qquad (7.3.1)$$

Substituting equation (7.3.1) into equation (7.2.2), we obtain the constitutive relation for an isotropic medium in the following form:

$$\mathbf{T}(\mathbf{R}) = \lambda\,[\nabla \cdot \mathbf{u}(\mathbf{R})]\mathbf{I} + \mu[\nabla\mathbf{u}(\mathbf{R}) + \mathbf{u}(\mathbf{R})\nabla] \qquad (7.3.2)$$

where $\mathbf{T}$ represents the stress tensor, $\mathbf{u}$ is the displacement vector, and $\mathbf{I}$ stands for the idemfactor. Across a surface containing the point $\mathbf{R}$, the tensor $\mathbf{T}$ is related to the traction $\mathbf{t}$ at the surface as follows:

164   *Seismic methods*

$$\mathbf{t} = \mathbf{n} \cdot \mathbf{T} = \mathbf{T} \cdot \mathbf{n} \qquad (7.3.3)$$

where $\mathbf{n}$ is the normal unit vector of the surface. The equation of wave motion in the frequency domain with time dependence $\exp(i\omega t)$ is

$$\nabla \cdot \mathbf{T}(\mathbf{R}) + \rho\omega^2 \mathbf{u}(\mathbf{R}) = -\rho \mathbf{f}(\mathbf{R}) \qquad (7.3.4)$$

where ρ is the density and $\mathbf{f}$ is the body force per unit mass. Eliminating $\mathbf{T}$ from equations (7.3.2) and (7.3.4), we obtain for the displacement vector $\mathbf{u}$:

$$[(\lambda + \mu)\nabla\nabla \cdot + \mu\nabla^2]\mathbf{u}(\mathbf{R}) + \rho\omega^2 \mathbf{u}(\mathbf{R}) = -\rho \mathbf{f}(\mathbf{R}) \qquad (7.3.5)$$

The wave field described by this equation is composed of a P wave and an S wave. The P and S waves propagate with characteristic velocities.

To express the wave field defined by equations (7.3.2) and (7.3.4) in an integral form, we construct Green's displacement tensor $\mathbf{G}(\mathbf{R}, \mathbf{R}_0)$ and Green's stress tensor $\Sigma(\mathbf{R}, \mathbf{R}_0)$. $\mathbf{G}$ represents the displacement field at $\mathbf{R}$ generated by three mutually perpendicular concentrated forces applied at $\mathbf{R}_0$, and Σ represents the stress field at $\mathbf{R}$ generated by three mutually perpendicular concentrated forces at $\mathbf{R}_0$. $\mathbf{G}$ and Σ satisfy the same relations as $\mathbf{u}$ and $\mathbf{T}$ so that

$$\Sigma(\mathbf{R}, \mathbf{R}_0) = \lambda [\nabla \cdot \mathbf{G}(\mathbf{R}, \mathbf{R}_0)]\mathbf{I} + \mu[\nabla \mathbf{G}(\mathbf{R}, \mathbf{R}_0) + \mathbf{G}(\mathbf{R}, \mathbf{R}_0)\nabla] \qquad (7.3.6)$$

$$\nabla \cdot \Sigma(\mathbf{R}, \mathbf{R}_0) + \rho\omega^2 \mathbf{G}(\mathbf{R}, \mathbf{R}_0) = -\delta(\mathbf{R} - \mathbf{R}_0)\mathbf{I} \qquad (7.3.7)$$

Combining equations (7.3.6) and (7.3.7), we obtain for $\mathbf{G}$, in a manner analogous to that of equation (7.3.5):

$$[(\lambda + \mu)\nabla\nabla \cdot + \mu\nabla^2]\mathbf{G}(\mathbf{R}, \mathbf{R}_0) + \rho\omega^2 \mathbf{G}(\mathbf{R}, \mathbf{R}_0) = -\delta(\mathbf{R} - \mathbf{R}_0)\mathbf{I} \qquad (7.3.8)$$

Green's tensors $\mathbf{G}$ and Σ of the integral representations for the wave field can be obtained by solving for $\mathbf{G}$ from equation (7.3.8), after which Σ can be calculated using equation (7.3.6).

For a homogeneous whole-space, the solution of equation (7.3.8) can be written in the following form:

$$\mathbf{G}(\mathbf{R}, \mathbf{R}_0) = \frac{1}{\rho\omega^2} \{k_S^2 \mathbf{I} g_S(\mathbf{R}, \mathbf{R}_0) + \nabla[g_P(\mathbf{R}, \mathbf{R}_0) - g_S(\mathbf{R}, \mathbf{R}_0)]\nabla_0\} \qquad (7.3.9)$$

where g_P and g_S are known as the **free-space Green's functions** for the P and S waves, respectively. g_P and g_S satisfy the following scalar wave equations:

$$(\nabla^2 + k_P^2) g_P(\mathbf{R}, \mathbf{R}_0) = -\delta(\mathbf{R} - \mathbf{R}_0) \qquad (7.3.10)$$

$$(\nabla^2 + k_S^2) g_S(\mathbf{R}, \mathbf{R}_0) = -\delta(\mathbf{R} - \mathbf{R}_0) \qquad (7.3.11)$$

Wavenumbers k_P and k_S are defined by $k_P = \omega/c_P$ and $k_S = \omega/c_S$, with the propagation velocities c_P and c_S given by:

$$c_P = \left\{\frac{\lambda + 2\mu}{\rho}\right\}^{\frac{1}{2}} \tag{7.3.12}$$

$$c_S = \left\{\frac{\mu}{\rho}\right\}^{\frac{1}{2}} \tag{7.3.13}$$

The whole-space solutions for equations (7.3.10) and (7.3.11) are in three dimensions:

$$g_P(\mathbf{R}, \mathbf{R}_0) = \frac{e^{-ik_P R}}{4\pi R} \tag{7.3.14}$$

$$g_S(\mathbf{R}, \mathbf{R}_0) = \frac{e^{-ik_S R}}{4\pi R} \tag{7.3.15}$$

where $R = |\mathbf{R} - \mathbf{R}_0|$.

For a two dimensional model,

$$g_P(\mathbf{r}, \mathbf{r}_0) = -\frac{i}{4} H_0^{(2)}(k_P r) \tag{7.3.16}$$

$$g_S(\mathbf{r}, \mathbf{r}_0) = -\frac{i}{4} H_0^{(2)}(k_S r) \tag{7.3.17}$$

where $H_0^{(2)}$ is the zero-order Hankel function of the second kind, and $r = |\mathbf{r} - \mathbf{r}_0|$.

To convert equations (7.3.2)–(7.3.8) into integral representations for $\mathbf{u}$ at the point $\mathbf{R}$, we take the scalar product of equation (7.3.4) on the right by $\mathbf{G}$ and of equation (7.3.7) on the left by $\mathbf{u}$. Subtracting one of the equations thus obtained from the other, we get

$$(\nabla \cdot \mathbf{T}) \cdot \mathbf{G} - \mathbf{u} \cdot (\nabla \cdot \Sigma) = -\rho \mathbf{f} \cdot \mathbf{G} + \mathbf{u}\delta(\mathbf{R} - \mathbf{R}_0) \tag{7.3.18}$$

which can be written as follows:

$$\nabla \cdot (\mathbf{T} \cdot \mathbf{G} - \mathbf{u} \cdot \Sigma) = -\rho \mathbf{f} \cdot \mathbf{G} + \mathbf{u}\delta(\mathbf{R} - \mathbf{R}_0) \tag{7.3.19}$$

Next we make the substitution

$$\mathbf{Q} = \mathbf{T} \cdot \mathbf{G} - \mathbf{u} \cdot \Sigma \tag{7.3.20}$$

in Gauss's theorem (equation (7.2.11)) and apply the theorem to surface S_1 and region V_1 interior to it in the earth model in Fig. 7.1. Substituting equation (7.3.3) and (7.3.19) and taking into consideration the symmetry relations of the Green's tensors, equations (7.2.8) and (7.2.9), we obtain:

166 *Seismic methods*

$$\int_{S_1} \{\mathbf{t}(\mathbf{R}_0) \cdot \mathbf{G}(\mathbf{R}, \mathbf{R}_0) - \mathbf{u}(\mathbf{R}_0) \cdot [\mathbf{n}_0 \cdot \Sigma(\mathbf{R}, \mathbf{R}_0)]\} \mathrm{d}S_0$$
$$+ \int_{V_1} \rho \mathbf{f}(\mathbf{R}_0) \cdot \mathbf{G}(\mathbf{R}, \mathbf{R}_0) \mathrm{d}V_0 = \begin{cases} \mathbf{u}(\mathbf{R}), & \mathbf{R} \in V_1 \\ 0, & \mathbf{R} \in V_2 \end{cases} \quad (7.3.21)$$

where the density and Lamé's constants are those of region V_1.

To obtain the integral formula for a stress field in the region external to S_1, we apply equations (7.2.11) and (7.3.20) to region V_2 and the surfaces S_1 and S_2 that internally and externally bound V_2 (Fig. 7.1), while letting S_2 approach infinity. In order to calculate the contribution of the surface integral over S_2 at $\mathbf{R}$ as S_2 goes to infinity, we impose the radiation conditions on the far-field behaviour of $\mathbf{u}$ and $\mathbf{t}$. These conditions require that there are no sources at infinity, that is, that the waves propagate outward through S_2 as R goes to infinity. This condition is satisfied by making the integrand of the surface integral over S_2 vanish uniformly for all $\mathbf{R}_0$. Assuming that S_2 is a sphere with radius R and denoting the unit vector of R by $\mathbf{R}^0$, we obtain the frequency-domain radiation conditions as follows:

$$\lim_{R \to \infty} R\,[-\,\mathrm{i}k_\mathrm{P}\rho c_\mathrm{P}^2 \mathbf{u}(R\mathbf{R}^0) + \mathbf{t}(R\mathbf{R}^0)] \cdot \mathbf{R}^0 = 0 \quad \text{for all } \mathbf{R}^0 \quad (7.3.22)$$

$$\lim_{R \to \infty} |R\,[-\,\mathrm{i}k_\mathrm{S}\rho c_\mathrm{S}^2 \mathbf{u}(R\mathbf{R}^0) + \mathbf{t}(R\mathbf{R}^0)] \cdot (\mathbf{I} - \mathbf{R}^0\mathbf{R}^0)| = 0 \quad \text{for all } \mathbf{R}^0 \quad (7.3.23)$$

$$\lim_{R \to \infty} |R\mathbf{u}(R\mathbf{R}^0)| < \chi, \ \chi > 0 \quad (7.3.24)$$

The radiation conditions (7.3.22)–(7.3.24) make the surface integral over S_2 in equation (7.2.11) vanish, and we thus obtain the integral formula for the wave field outside surface S_1:

$$\int_{S_1} \{\mathbf{u}(\mathbf{R}_0) \cdot [\mathbf{n}_0 \cdot \Sigma(\mathbf{R}, \mathbf{R}_0)] - \mathbf{t}(\mathbf{R}_0) \cdot \mathbf{G}(\mathbf{R}, \mathbf{R}_0)\} \mathrm{d}S_0$$
$$+ \int_{V_2} \rho \mathbf{f}(\mathbf{R}_0) \cdot \mathbf{G}(\mathbf{R}, \mathbf{R}_0) \mathrm{d}V_0 = \begin{cases} \mathbf{u}(\mathbf{R}), & \mathbf{R} \in V_2 \\ 0, & \mathbf{R} \in V_1 \end{cases} \quad (7.3.25)$$

The material constants in equation (7.3.25) are those of V_2. The surface S_1 is any arbitrary surface that encloses all sources; it may be, for example, a discontinuity surface of material properties.

Assume next that S_1 is actually the surface of discontinuity in material properties. Let the density and Lamé's constants be ρ_1, λ_1, μ_1 in V_1, and ρ_2, λ_2, μ_2 in V_2 (Fig. 7.1). Let V_2 contain primary sources that generate an incident wave $\mathbf{u}^\mathrm{i}$. As a result of the discontinuity in density and Lamé's constants, the incident wave is scattered by S_1. Denoting the scattered waves in V_2 by $\mathbf{u}^\mathrm{s}$, the total displacement outside S_1 can be written as follows:

$$\mathbf{u}(\mathbf{R}) = \mathbf{u}^\mathrm{i}(\mathbf{R}) + \mathbf{u}^\mathrm{s}(\mathbf{R}), \quad \mathbf{R} \in V_2 \quad (7.3.26)$$

Inside S_1 we denote the displacement vector by $\mathbf{u}^\mathrm{t}$, hence

Integral formulas for elastic wave fields in an isotropic medium 167

$$\mathbf{u}(\mathbf{R}) = \mathbf{u}^i(\mathbf{R}), \quad \mathbf{R} \in V_1 \qquad (7.3.27)$$

Applying equation (7.3.21) to the incident wave field in the external region V_2, we obtain:

$$\int_{S_1} \{\mathbf{t}^i(\mathbf{R}_0) \cdot \mathbf{G}^2(\mathbf{R}, \mathbf{R}_0) - \mathbf{u}^i(\mathbf{R}_0) \cdot [\mathbf{n}_0 \cdot \Sigma^2(\mathbf{R}, \mathbf{R}_0)]\} dS_0 = 0, \quad \mathbf{R} \in V_2 \qquad (7.3.28)$$

The sources of the scattered field $\mathbf{u}^s$ are on or inside S_1, and hence we can apply equation (7.3.25) outside S_1, without the body sources, and obtain:

$$\int_{S_1} \{\mathbf{u}^s(\mathbf{R}_0) \cdot [\mathbf{n}_0 \cdot \Sigma^2(\mathbf{R}, \mathbf{R}_0)] - \mathbf{t}^s(\mathbf{R}_0) \cdot \mathbf{G}^2(\mathbf{R}, \mathbf{R}_0)\} dS_0$$

$$= \mathbf{u}^s(\mathbf{R}), \quad \mathbf{R} \in V_2 \qquad (7.3.29)$$

Combining equations (7.3.28) and (7.3.29), and substituting equations (7.3.26), we obtain for the total wave field outside S_1:

$$\int_{S_1} \{\mathbf{u}(\mathbf{R}_0) \cdot [\mathbf{n}_0 \cdot \Sigma^2(\mathbf{R}, \mathbf{R}_0)] - \mathbf{t}(\mathbf{R}_0) \cdot \mathbf{G}^2(\mathbf{R}, \mathbf{R}_0)\} dS_0 + \mathbf{u}^i(\mathbf{R})$$

$$= \mathbf{u}(\mathbf{R}), \quad \mathbf{R} \in V_2 \qquad (7.3.30)$$

where $\mathbf{G}^2$ and Σ^2 are Green's displacement tensor and Green's stress tensor having medium constants ρ_2, λ_2, μ_2.

The integral formula for the wave field inside surface S_1 can be derived using equation (7.3.21). Remembering that the primary sources are outside S_1, we obtain:

$$\int_{S_1} \{\mathbf{t}(\mathbf{R}_0) \cdot \mathbf{G}^1(\mathbf{R}, \mathbf{R}_0) - \mathbf{u}(\mathbf{R}_0) \cdot [\mathbf{n}_0 \cdot \Sigma^1(\mathbf{R}, \mathbf{R}_0)]\} dS_0 = \mathbf{u}^i(\mathbf{R}), \quad \mathbf{R} \in V_1 \qquad (7.3.31)$$

where $\mathbf{G}^1$ and Σ^1 denote Green's displacement tensor and Green's stress tensor with material constants ρ_1, λ_1, μ_1. Vectors $\mathbf{u}(\mathbf{R}_0)$ and $\mathbf{t}(\mathbf{R}_0)$ on S_1 are the same as those in equation (7.3.30).

If the values of $\mathbf{u}$ and $\mathbf{t}$ at S_1 are known, the total displacement can be calculated in V_2 from equation (7.3.30) and in V_1 from equation (7.3.31). However, for most practical problems, either $\mathbf{u}$ or $\mathbf{t}$ or both are unknown. When one of them is known, the other can be derived from an integral equation obtained by moving the point $\mathbf{R}$ in equation (7.3.30) to surface S_1. When both the surface sources are unknown, a system of integral equations for $\mathbf{u}$ and $\mathbf{t}$ can be formulated from equations (7.3.30) and (7.3.31) by moving $\mathbf{R}$ on to surface S_1. The integral equations can be solved by using standard numerical methods.

The above discussion of wave fields was concerned with the frequency domain. The time-domain formulas can be obtained from the corresponding frequency-domain formulas by taking the inverse Fourier transform of the defining functions of the problem.

7.4 SEPARATION OF ELASTIC WAVE FIELDS INTO A COMPRESSIONAL AND A ROTATIONAL MODE

In this section we consider how an elastic wave field is separated into a compressional (P) and a rotational (S) wave mode. We know from vector analysis that any vector field that vanishes at infinity can be expressed in terms of the gradient of a scalar potential and the curl of a divergence-free vector potential.

The equation satisfied by displacement vector **u** in the absence of body forces can be rewritten from equation (7.3.5) as follows:

$$[(\lambda + \mu)\nabla\nabla \cdot + \mu\nabla^2]\mathbf{u} + \rho\omega^2\mathbf{u} = 0 \tag{7.4.1}$$

Substituting $\nabla^2 = \nabla\nabla \cdot + \nabla \times \nabla \times$, equation (7.4.1) takes the following form:

$$(\lambda + 2\mu)\nabla\nabla \cdot \mathbf{u} - \mu\nabla \times \nabla \times \mathbf{u} + \rho\omega^2\mathbf{u} = 0 \tag{7.4.2}$$

In order to specify the compressional and the rotational wave modes in equation (7.4.2), we define the Lamé potentials φ and **F** as follows:

$$\mathbf{u} = \nabla\varphi + \nabla \times \mathbf{F} \tag{7.4.3a}$$

$$\nabla \cdot \mathbf{F} = 0 \tag{7.4.3b}$$

Taking separately the divergence and the curl of equation (7.4.3a) and applying equation (7.4.3b), we obtain:

$$\nabla \cdot \mathbf{u} = \nabla^2\varphi \tag{7.4.4}$$

$$\nabla \times \mathbf{u} = -\nabla^2\mathbf{F} \tag{7.4.5}$$

Substitution of equations (7.4.4) and (7.4.5) into equation (7.4.2) yields two independent equations for P and S waves, respectively:

$$\nabla^2\varphi + k_P^2\varphi = 0 \tag{7.4.6}$$

$$\nabla^2\mathbf{F} + k_S^2\mathbf{F} = 0 \tag{7.4.7}$$

where the wavenumbers are $k_P = \omega/c_P$ and $k_S = \omega/c_S$. The propagation velocities c_P and c_S are given by equations (7.3.12) and (7.3.13), respectively. Substituting equation (7.4.4) into equation (7.4.6), and equation (7.4.5) into equation (7.4.7), the scalar potential φ and the vector potential **F** are expressed in terms of the displacement vector **u** as follows:

$$\varphi = -\frac{1}{k_P^2}\nabla \cdot \mathbf{u} \tag{7.4.8}$$

$$\mathbf{F} = \frac{1}{k_S^2}\nabla \times \mathbf{u} \tag{7.4.9}$$

Taking into consideration that $\nabla \cdot \mathbf{u}$ is the change in volume and that $\nabla \times \mathbf{u}$ represents the rotation of a material particle, we note, referring to equations (7.4.8) and (7.4.9), that equation (7.4.6) represents a compressional (P) wave field, and equation (7.4.7) a rotational (S) wave field.

In conventional seismic surveys P waves are of major importance, since they carry most of the seismic energy generated by the conventional seismic sources. This makes it possible to apply an approximation method which involves treating the elastic medium as purely acoustic. Accordingly, it is required that the rotational waves represented by equation (7.4.7) vanish so that the wave field is a purely compressional one represented by equation (7.4.6). In the rest of this chapter we consider the Helmholtz and Kirchhoff integral formulas that are obtained by conversion from the acoustic wave equation, equation (7.4.6).

7.5 INTEGRAL FORMULAS FOR ACOUSTIC WAVE FIELDS IN THE FREQUENCY DOMAIN

Consider a region V enclosed by a surface S, and let V be filled with an acoustic medium characterized by propagation velocity c_P of a compressional wave. The wave motion in V is described by the following scalar wave equation:

$$\nabla^2 \varphi(\mathbf{R}) + k_P^2 \varphi(\mathbf{R}) = -f(\mathbf{R}) \qquad (7.5.1)$$

where f describes the source density, and φ is the displacement potential. Potential φ is given in terms of the displacement vector $\mathbf{u}$ as follows:

$$\mathbf{u} = \nabla \varphi \qquad (7.5.2)$$

$$\varphi = -\frac{1}{k_P^2} \nabla \cdot \mathbf{u} \qquad (7.5.3)$$

The wavenumber is $k_P = \omega/c_P$, and the propagation velocity c_P of the P wave is given by equation (7.3.12).

In order to solve equation (7.5.1) in the integral form we define a Green's function $g(\mathbf{R}, \mathbf{R}_0)$ which, for a forward problem, satisfies the following wave equation:

$$\nabla^2 g(\mathbf{R}, \mathbf{R}_0) + k_P^2 g(\mathbf{R}, \mathbf{R}_0) = -\delta(\mathbf{R} - \mathbf{R}_0) \qquad (7.5.4)$$

For the forward problem, the frequency-domain Green's function $g(\mathbf{R}, \mathbf{R}_0)$ and the corresponding time-domain Green's function, in which time flows in the positive direction, are the Fourier transforms of each other. The solution of equation (7.5.4) satisfies boundary conditions which are dependent upon the earth model under consideration. The far-field behaviour is imposed by requiring that the waves travel outward through a surface S_∞ as the surface recedes to infinity. This physical behaviour is satisfied if the potential φ and Green's function g satisfy the following radiation conditions:

$$\lim_{R \to \infty} (\mathbf{R}^0 \cdot \nabla\varphi - ik_P\varphi) = 0 \tag{7.5.5}$$

$$\lim_{R \to \infty} (\mathbf{R}^0 \cdot \nabla g - ik_P g) = 0 \tag{7.5.6}$$

where R is the radius from the calculation point $\mathbf{R}$ to surface S_∞, and $\mathbf{R}^0$ is the unit vector of R.

For an inverse problem, the frequency-domain Green's function $g(\mathbf{R}_0, \mathbf{R})$, corresponding to the time-domain Green's function in which time flows in the negative direction, is defined as follows:

$$\nabla^2 g(\mathbf{R}_0, \mathbf{R}) + k_P^2 g(\mathbf{R}_0, \mathbf{R}) = -\delta(\mathbf{R}_0 - \mathbf{R}) \tag{7.5.7}$$

Green's functions for the forward and the inverse problems satisfy the symmetry relation

$$g(\mathbf{R}_0, \mathbf{R}) = g^*(\mathbf{R}, \mathbf{R}_0) \tag{7.5.8}$$

where the asterisk denotes the complex conjugate.

Next we substitute functions φ and g into Green's second identity for scalars and apply it to the earth model:

$$\int_V (g\nabla^2\varphi - \varphi\nabla^2 g)dV = \int_S (g\nabla\varphi - \varphi\nabla g) \cdot \mathbf{n}dS \tag{7.5.9}$$

where $\mathbf{n}$ is the outward unit vector of surface S. Substituting equation (7.5.1) first with equation (7.5.4) and then with equation (7.5.7) into equation (7.5.9), and applying the symmetry relation (7.5.8), we obtain [71]:

$$\varphi(\mathbf{R}) = \int_V f(\mathbf{R}_0)g(\mathbf{R}, \mathbf{R}_0)dV_0 + \int_S [g(\mathbf{R}, \mathbf{R}_0)\nabla_0\varphi(\mathbf{R}_0)$$
$$- \varphi(\mathbf{R}_0)\nabla_0 g(\mathbf{R}, \mathbf{R}_0)] \cdot \mathbf{n}_0 dS_0 \tag{7.5.10}$$

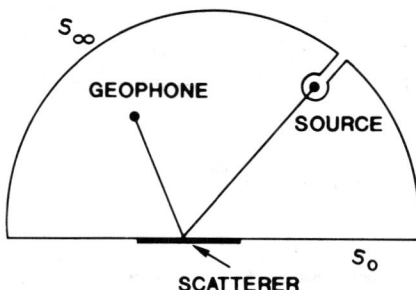

Figure 7.2 Closed integration surface for a seismic experiment. The hemispherical portion of the surface is at a great distance from the source. After [72].

Integral formulas for acoustic wave fields in the frequency domain 171

$$\varphi(\mathbf{R}_0) = \int_V f(\mathbf{R}) g(\mathbf{R}_0, \mathbf{R}) dV + \int_S [g(\mathbf{R}_0, \mathbf{R}) \nabla \varphi(\mathbf{R})$$
$$- \varphi(\mathbf{R}) \nabla g(\mathbf{R}_0, \mathbf{R})] \cdot \mathbf{n} dS \qquad (7.5.11)$$

Equations (7.5.10) and (7.5.11) represent the displacement potential at points $\mathbf{R}$ and $\mathbf{R}_0$ due to the source function f in region V, and the potential φ and its normal gradient $\mathbf{n} \cdot \nabla \varphi$ at surface S. If the surface containing the secondary sources can be simulated by a plane, it is practical to consider the closed surface S as being composed of a plane S_0 and a large hemisphere S_∞ (see Fig. 7.2). For forward modelling problems, S_0 may coincide with a reflecting surface; for inverse problems, S_0 may represent the measurement surface, which is usually the ground surface. If we allow S_∞ go to infinity and apply the radiation conditions, equations (7.5.5) and (7.5.6), the surface integrals over S_∞ in equations (7.5.10) and (7.5.11) vanish, and we are left with the integrals over S_0. Assuming that the volume sources lie outside S, and that S_0 is excited by external sources, we obtain two integral formulas known as the **Helmholtz construction** and **reconstruction integrals**:

$$\varphi(\mathbf{R}) = \int_{S_0} \left[g(\mathbf{R}, \mathbf{R}_0) \frac{\partial \varphi(\mathbf{R}_0)}{\partial n_0} - \varphi(\mathbf{R}_0) \frac{\partial g(\mathbf{R}, \mathbf{R}_0)}{\partial n_0} \right] dS_0 \qquad (7.5.12)$$

$$\varphi(\mathbf{R}_0) = \int_{S_0} \left[g(\mathbf{R}_0, \mathbf{R}) \frac{\partial \varphi(\mathbf{R})}{\partial n} - \varphi(\mathbf{R}) \frac{\partial g(\mathbf{R}_0, \mathbf{R})}{\partial n} \right] dS \qquad (7.5.13)$$

For a non-homogeneous subsurface, Green's function g may be rather complicated. For seismic surveys typical of those conducted in applied geophysics, Green's function for a plane-layered earth with the source in the top layer is of particular interest. Green's function must be specified separately for each particular earth model being investigated. In this book we are mainly concerned with the theoretical basis of geophysical modelling, so that, only the simplest types of Green's functions will be considered, such as the wholespace Green's function and the half-space Green's functions having either free surface or rigid surface boundary conditions.

The free-space Green's function is the particular solution of equation (7.5.4) which satisfies radiation condition (7.5.6) but no other boundary conditions. Given these conditions, the free-space Green's function g_0 takes the following form:

$$g_0(\mathbf{R}, \mathbf{R}_0) = \frac{e^{-ik_P R}}{4\pi R} \qquad (7.5.14a)$$

$$g_0(\mathbf{R}_0, \mathbf{R}) = \frac{e^{ik_P R}}{4\pi R} \qquad (7.5.14b)$$

with $R = |\mathbf{R} - \mathbf{R}_0|$, and $k_P^2 = \omega/c_P$.

172 Seismic methods

Substituting equations (7.5.14) into the Helmholtz integrals, equations (7.5.12) and (7.5.13), we obtain:

$$\varphi(\mathbf{R}) = \int_{S_0} \frac{e^{-ik_P R}}{4\pi} \left\{ \frac{1}{R} \frac{\partial \varphi(\mathbf{R}_0)}{\partial n_0} - \varphi(\mathbf{R}_0) \frac{\partial}{\partial n_0}\left(\frac{1}{R}\right) + \varphi(\mathbf{R}_0) \frac{ik_P}{R} \frac{\partial R}{\partial n_0} \right\} dS_0 \quad (7.5.15)$$

$$\varphi(\mathbf{R}_0) = \int_{S_0} \frac{e^{ik_P R}}{4\pi} \left\{ \frac{1}{R} \frac{\partial \varphi(\mathbf{R})}{\partial n} - \varphi(\mathbf{R}) \frac{\partial}{\partial n}\left(\frac{1}{R}\right) - \varphi(\mathbf{R}) \frac{ik_P}{R} \frac{\partial R}{\partial n} \right\} dS \quad (7.5.16)$$

In Section 7.6 we will derive, using equation (7.5.15), the Kirchhoff and Rayleigh–Sommerfeld time-domain integral formulas for forward seismic modelling. The corresponding formulas for the inverse modelling procedure may be obtained from equation (7.5.16) by a similar route.

7.6 INTEGRAL FORMULAS FOR ACOUSTIC WAVE FIELDS IN THE TIME DOMAIN

The wave equation for the displacement potential Φ in the time domain can be expressed by the inverse Fourier transform of equation (7.5.1):

$$\nabla^2 \Phi(\mathbf{R}; t) - \frac{1}{c_P^2} \frac{\partial^2 \Phi(\mathbf{R}; t)}{\partial t^2} = -F(\mathbf{R}; t) \quad (7.6.1)$$

where the time-dependent functions Φ and F are related to the corresponding frequency dependent functions φ and f by the inverse Fourier transform:

$$A(t) = \frac{1}{2\pi} \int_{-\infty}^{\infty} a(\omega) e^{i\omega t} d\omega \quad (7.6.2)$$

where $A(t)$ denotes the functions $\Phi(t)$ and $F(t)$, and $a(\omega)$ the functions $\varphi(\omega)$ and $f(\omega)$.

The wave equation for Green's function G in the time domain is, by analogy with equation (7.6.1):

$$\nabla^2 G(\mathbf{R}, \mathbf{R}_0; t, t_0) - \frac{1}{c_P^2} \frac{\partial^2 G(\mathbf{R}, \mathbf{R}_0; t, t_0)}{\partial t^2} = -\delta(\mathbf{R} - \mathbf{R}_0)\delta(t - t_0) \quad (7.6.3)$$

The wave field defined by equations (7.6.1) and (7.6.3) could be converted into an integral representation by substituting functions Φ and G into Green's second identity, equation (7.5.9). However, we shall derive the time-domain integral formulas directly from the corresponding frequency-domain formulas considered in Section 7.5. Applying the inverse Fourier transform equation (7.6.2) to equation (7.6.15) and interchanging the order of integration, we obtain the Kirchhoff formula [71]:

$$\Phi(\mathbf{R}; t) = \frac{1}{4\pi} \int_{S_0} \left\{ \frac{1}{R} \frac{\partial \Phi(\mathbf{R}_0; \tau_-)}{\partial n_0} - \Phi(\mathbf{R}_0; \tau_-) \frac{\partial}{\partial n_0} \left(\frac{1}{R} \right) \right.$$

$$\left. + \frac{1}{c_P R} \frac{\partial R}{\partial n_0} \frac{\partial \Phi(\mathbf{R}_0; \tau_-)}{\partial t_0} \right\} dS_0 \qquad (7.6.4)$$

where $\tau_- = t - R/c_P$.

Note that the Green's function applied in equation (7.6.4) is the free-space Green's function G_0, which can be obtained from equations (7.5.14a) and (7.6.2):

$$G_0 = \frac{\delta(t - t_0 - R/c_P)}{4\pi R} \qquad (7.6.5)$$

The difficulty in applying the Kirchhoff formula (7.6.4) is that to calculate Φ requires the simultaneous imposition of boundary values on both Φ and its normal derivative. Thus, for the Kirchhoff formula, the boundary conditions overspecify the problem.

Assume next that surface S_0 coincides with a material discontinuity, such as the earth–air interface, for which the reflection coefficient $r = -1$. At such an interface, the potential Φ and Green's function G satisfy the **Dirichlet** or **free-surface** boundary conditions:

$$\Phi_- = 0 \qquad (7.6.6a)$$

$$\partial \Phi_-/\partial n_0 = 2\partial \Phi_0/\partial n_0 \qquad (7.6.6b)$$

$$G_- = 0 \qquad (7.6.7a)$$

$$\partial G_-/\partial n_0 = 2\partial G_0/\partial n_0 \qquad (7.6.7b)$$

Substituting equations (7.6.7), Green's function G_- takes the form:

$$G_- = \frac{\delta(t - t_0 - R/c_P)}{4\pi R} - \frac{\delta(t - t_0 - R'/c_P)}{4\pi R'} \qquad (7.6.8)$$

where $R' = |\mathbf{R} - \mathbf{R}_0'|$, and $\mathbf{R}_0'$ is the mirror image of $\mathbf{R}_0$ relative to plane S_0. Substituting the free-surface Green's function, equation (7.6.8) into equation (7.6.4), we obtain:

$$\Phi_-(\mathbf{R}; t) = \frac{1}{2\pi} \int_{S_0} \left\{ \frac{1}{c_P R} \frac{\partial R}{\partial n_0} \frac{\partial \Phi(\mathbf{R}_0; \tau_-)}{\partial t} - \Phi(\mathbf{R}_0; \tau_-) \frac{\partial}{\partial n_0} \left(\frac{1}{R} \right) \right\} dS_0 \qquad (7.6.9)$$

If the reflection coefficient at S_0 is $r = +1$, Φ and G satisfy the **Neumann** or **rigid-surface** boundary conditions:

$$\Phi_+ = 2\Phi_0 \qquad (7.6.10a)$$

$$\partial \Phi_+/\partial n_0 = 0 \qquad (7.6.10b)$$

174   Seismic methods

$$G_+ = 2G_0 \quad (7.6.11a)$$

$$\partial G_+/\partial n_0 = 0 \quad (7.6.11b)$$

at interface S_0. The Green's function satisfying equations (7.6.11) can be written as follows:

$$G_+ = \frac{\delta(t - t_0 - R/c_P)}{4\pi R} + \frac{\delta(t - t_0 - R'/c_P)}{4\pi R'} \quad (7.6.12)$$

where R' is as in equation (7.6.8). Substituting the rigid-surface Green's function, equation (7.6.12), into equation (7.6.4), we obtain:

$$\Phi_+(\mathbf{R}; t) = \frac{1}{2\pi} \int_{S_0} \frac{1}{R} \frac{\partial \Phi(\mathbf{R}_0; \tau_-)}{\partial n_0} dS_0 \quad (7.6.13)$$

Equations (7.6.9) and (7.6.13) are known as the **Rayleigh–Sommerfeld integral formulas**, which have the advantage, in contrast to the Kirchhoff formulas, of being self-consistent, that is, the need for simultaneous imposition of boundary values on both the potential Φ and its normal derivative does not arise. The free-surface result has been most widely used in migration problems, whereas the applications of the rigid-surface result are more typically associated with scattering problems.

7.7 APPLICATIONS

In this section, we shall briefly discuss the application of Helmholtz and Kirchhoff integral formulas in both forward and inverse seismic modelling problems. First, we derive a simple formula for calculating the displacement potential of an acoustic wave field scattered at a region of discontinuity within the subsurface. Second, we derive a formula for estimating reflection coefficients in the subsurface by migration of seismic data.

7.7.1 Scattering

Assume that the medium surrounding scatterer A is homogeneous, which makes it possible to use the free-space Green's function g_0 as given by equation (7.6.14a). Substituting g_0 into the Helmholtz construction integral, equation (7.5.12), we obtain for the displacement potential φ at the measurement point $\mathbf{R}$ (see Fig. 7.2):

$$\varphi(\mathbf{R}) = \frac{1}{4\pi} \int_{S_0} \left[\frac{e^{-ik_P R}}{R} \frac{\partial \varphi(\mathbf{R}_0)}{\partial n_0} - \varphi(\mathbf{R}_0) \frac{\partial}{\partial n_0}\left(\frac{e^{-ik_P R}}{R}\right) \right] dS_0 \quad (7.7.1)$$

where $R = |\mathbf{R} - \mathbf{R}_0|$, and $k_P = \omega/c_P$.

Let the primary source be a point source located at $\mathbf{R}_S$, which generates a spherical incident wave whose potential at point $\mathbf{R}_0$ of the scatterer can be expressed as follows [72; 73]:

$$\varphi_1(\mathbf{R}_0) = \frac{e^{-ikR_S}}{R_S} \varphi_0 \qquad (7.7.2)$$

where φ_0 is the source strength, and $R_S = |\mathbf{R}_0 - \mathbf{R}_S|$.

We assume next that the problem satisfies the weak scattering condition, namely that the reflection coefficient r at A is so small that multiple scattering can be neglected. We further assume that the reflection angles are small, due to which condition r can be considered to be constant. The potential of the scattered wave, φ_A, can be expressed in terms of r and φ_1 as follows:

$$\varphi_A(\mathbf{R}_0) = r\varphi_1(\mathbf{R}_0) \qquad (7.7.3)$$

on scatterer A; it is zero on other parts of S_0. Since φ_A is an upward propagating wave, it is an image of $r\varphi_1$ in the vicinity of A, hence:

$$\frac{\partial \varphi_A}{\partial n} = -r \frac{\partial \varphi_1}{\partial n} \qquad (7.7.4)$$

Substituting equations (7.7.2), (7.7.3) and (7.7.4) into equation (7.7.1), we obtain the displacement potential of the scattered wave at the observation point $\mathbf{R}$ in terms of the source strength and reflection coefficient:

$$\varphi_A(\mathbf{R}) = -\frac{r\varphi_0}{4\pi} \int_A \frac{\partial}{\partial n_0} \left(\frac{e^{-ik_P(R_S + R)}}{R_S R} \right) dA_0 \qquad (7.7.5)$$

The potential φ_A is calculated by using equation (7.7.5) for a sufficient set of ω values, and the results are then used to calculate, by inverse Fourier transformation, the time-domain potential of the scattered wave at the precise instant of interest.

7.7.2 Migration

The seismic inversion method, known as migration, is a method by which seismic data are used for reconstructing the structures of subsurface reflectors. Thus, the aim of migration is to solve the problem considered in Section 7.7.1 in reverse order or, in other words, to determine the reflection coefficient at each point of the subsurface on the basis of the known wave field at the ground surface [72; 74]. Here we consider the problem as Kirchhoff migration, which is based on the inverse formula of equation (7.6.4), that is:

$$\Phi(\mathbf{R}_0; t_0) = \frac{1}{4\pi} \int_{S_0} \left\{ \frac{1}{R} \frac{\partial \Phi(\mathbf{R}; \tau_+)}{\partial n} - \Phi(\mathbf{R}; \tau_+) \frac{\partial}{\partial n}\left(\frac{1}{R}\right) \right.$$

$$\left. + \frac{1}{c_P R} \frac{\partial R}{\partial n} \frac{\partial \Phi(\mathbf{R}; \tau_+)}{\partial t} \right\} dS \qquad (7.7.6)$$

where $\tau_+ = t_0 + R/c_P$, $\mathbf{R}_0$ is a migration point on the subsurface, and $\mathbf{R}$ is a point on the ground surface S_0 containing the measured data. The application of the

Kirchhoff integral to the free-space Green's function again means that the condition of weak scattering is assumed to be valid, so that multiple reflections are neglected.

Since the upward propagating reflected wave field and its normal derivative on the ground surface S_0 are both known, we can calculate the wave field on the subsurface by using equation (7.7.6). We are interested in $\Phi(\mathbf{R}_0; t_0)$ only at the particular instant of time at which the wave measured at $\mathbf{R}$ may have been reflected by a discontinuity located at $\mathbf{R}_0$. Consequently, we model the value of $\Phi(\mathbf{R}_0)$ at the instant $t_S = R_S/c_P$, where R_S is the distance of point $\mathbf{R}_0$ from the source point $\mathbf{R}_S$.

The effect of the earth's surface on the scattered wave is taken into consideration by applying the free-surface boundary conditions (7.6.6), at S_0. Let S_0 coincide with the plane $z = 0$ of a Cartesian system of coordinates. Applying the free-surface Green's function, equation (7.6.8), equation (7.7.6) simplifies to:

$$\Phi(\mathbf{R}_0; t_S) = -\frac{1}{2\pi} \int_{z=0} \frac{1}{R} u_z(\mathbf{R}; \tau_+) \, dS \qquad (7.7.7)$$

where $u_z = -\partial \Phi/\partial z$ is the z component of displacement vector $\mathbf{u}$ at the earth's surface. Since Φ is the upward propagating reflected wave, we note, in view of equation (7.6.6b), that the vertical displacement registered by the geophone at $z = 0$ equals $2u_z$.

Equation (7.7.7) represents the signals observed by a buried geophone at the time required by the wave field to propagate from the source to the geophone. This is the time at which a reflection would arrive at the buried geophone from a reflector exactly below the geophone. Subsequent division by the amplitude of the direct wave at each subsurface point yields, according to equation (7.7.3), approximate reflection coefficients.

Appendix A

Green's function for scalar potential in a two-layer half-space

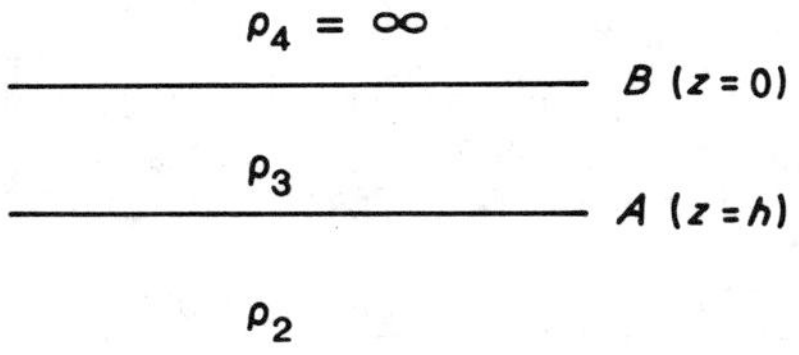

$\rho_4 = \infty$

—————————————————— $B \ (z = 0)$

ρ_3

—————————————————— $A \ (z = h)$

ρ_2

Figure A.1 Two-layer, half-space earth model.

A two-layer half-space is shown in Fig. A.1, in which the ground surface B is at $z = 0$, and the horizontal boundary A between regions 2 and 3, of resistivities ρ_2 and ρ_3, is at $z = h$. Let us denote Green's function for regions 2 and 3 of this space by G_{ab}, $a = 2, 3$, $b = 2, 3$ where a describes the region which contains the calculation point $\mathbf{R} = (x, y, z)$, and b the region which contains the source point $\mathbf{R}_0 = (x_0, y_0, z_0)$. Green's function satisfies the partial differential equations

$$\nabla^2 G_{22} = -\delta(\mathbf{R} - \mathbf{R}_0)$$
$$\nabla^2 G_{32} = 0$$
$$\nabla^2 G_{33} = -\delta(\mathbf{R} - \mathbf{R}_0)$$
$$\nabla^2 G_{23} = 0 \qquad\qquad (A.1)$$

and the boundary conditions

$$G_{22} = G_{32}$$
$$G_{23} = G_{33}$$

$$\frac{1}{\rho_2}\frac{\partial G_{22}}{\partial z} = \frac{1}{\rho_3}\frac{\partial G_{32}}{\partial z}$$

$$\frac{1}{\rho_2}\frac{\partial G_{23}}{\partial z} = \frac{1}{\rho_3}\frac{\partial G_{33}}{\partial z} \tag{A.2}$$

on surface A, and

$$\frac{\partial G_{32}}{\partial z} = 0$$

$$\frac{\partial G_{33}}{\partial z} = 0 \tag{A.3}$$

on ground surface B.

Using the notation

$$R = |\mathbf{R} - \mathbf{R}_0|$$
$$r^2 = (x - x_0)^2 + (y - y_0)^2$$
$$K = \frac{\rho_3 - \rho_2}{\rho_3 + \rho_2}$$

we obtain for Green's functions G_{ab} [75]

$$G_{22} = \frac{1}{4\pi}\left\{\frac{1}{R} + K\sum_{n=0}^{\infty}\frac{(-K)^n}{[(2(n-1)h + z_0 + z)^2 + r^2]^{\frac{1}{2}}}\right.$$
$$\left. + \sum_{n=0}^{\infty}\frac{(-K)^n}{[(2nh + z_0 + z)^2 + r^2]^{\frac{1}{2}}}\right\} \tag{A.4}$$

$$G_{23} = \frac{1}{4\pi}(1-K)\left\{\sum_{n=0}^{\infty}\frac{(-K)^n}{[(2nh + z - z_0)^2 + r^2]^{\frac{1}{2}}}\right.$$
$$\left. + \sum_{n=0}^{\infty}\frac{(-K)^n}{[(2nh + z_0 + z)^2 + r^2]^{\frac{1}{2}}}\right\} \tag{A.5}$$

$$G_{32} = \frac{1}{4\pi}(1+K)\left\{\sum_{n=0}^{\infty}\frac{(-K)^n}{[(2nh + z_0 - z)^2 + r^2]^{\frac{1}{2}}}\right.$$
$$\left. + \sum_{n=0}^{\infty}\frac{(-K)^n}{[(2nh + z_0 + z)^2 + r^2]^{\frac{1}{2}}}\right\} \tag{A.6}$$

$$G_{33} = \frac{1}{4\pi}\left\{\frac{1}{R} - K\sum_{n=0}^{\infty}\frac{(-K)^n}{[(2(n+1)h+z-z_0)^2+r^2]^{\frac{1}{2}}}\right.$$

$$+ \sum_{n=0}^{\infty}\frac{(-K)^n}{[(2nh+z_0+z)^2+r^2]^{\frac{1}{2}}}$$

$$- K\sum_{n=0}^{\infty}\frac{(-K)^n}{[(2(n+1)h-z+z_0)^2+r^2]^{\frac{1}{2}}}$$

$$\left. - K\sum_{n=0}^{\infty}\frac{(-K)^n}{[(2(n+1)h-z_0-z)^2+r^2]^{\frac{1}{2}}}\right\} \quad (A.7)$$

Appendix B

Green's function for scalar potential in a half-space with a vertical contact

A half-space containing a vertical contact is shown in Fig. B.1, in which the contact A is represented by the plane $x = 0$, and the ground surface B by plane $z = 0$. The source point $\mathbf{R}_0$ is assumed to be located in region 2.

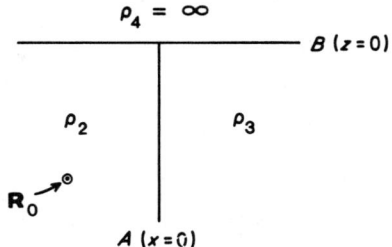

Figure B.1 Half-space with a vertical contact.

Green's functions G_2 and G_3, transmitting the potential to regions 2 and 3, satisfy the partial differential equations

$$\nabla^2 G_2 = -\delta(\mathbf{R} - \mathbf{R}_0)$$
$$\nabla^2 G_3 = 0 \tag{B.1}$$

and the boundary conditions

$$G_2 = G_3$$
$$\frac{1}{\rho_2} \frac{\partial G_2}{\partial x} = \frac{1}{\rho_3} \frac{\partial G_3}{\partial x} \tag{B.2}$$

on the vertical contact A, and

$$\frac{\partial G_2}{\partial z} = \frac{\partial G_3}{\partial z} = 0 \tag{B.3}$$

on ground surface B.

We obtain as the solution to this boundary-value problem the following Green's functions [22; 36]:

$$G_2 = \frac{1}{4\pi}\left\{\frac{1}{|\mathbf{R}-\mathbf{R}_0|} + \frac{1}{|\mathbf{R}-\mathbf{R}_0'|} + \frac{K}{|\mathbf{R}-\mathbf{R}_0''|} + \frac{K}{|\mathbf{R}-\mathbf{R}_0'''|}\right\} \tag{B.4}$$

$$G_3 = \frac{1+K}{4\pi}\left\{\frac{1}{|\mathbf{R}-\mathbf{R}_0|} + \frac{1}{|\mathbf{R}-\mathbf{R}_0'|}\right\} \tag{B.5}$$

where

$$\mathbf{R}_0 = (x_0, y_0, z_0)$$
$$\mathbf{R}_0' = (x_0, y_0, -z_0)$$
$$\mathbf{R}_0'' = (-x_0, y_0, z_0)$$
$$\mathbf{R}_0''' = (-x_0, y_0, -z_0)$$
$$k = \frac{\rho_3 - \rho_2}{\rho_3 + \rho_2}$$

Appendix C

Green's function for scalar potential in an anisotropic half-space

Consider Green's function $\mathscr{G}$ for an anisotropic half-space, where one of the principal axes of conductivity is parallel to the ground surface. The principal values of conductivity $\tilde{g}$ are g_a, g_b and g_c, and those of resistivity $\tilde{\rho}$ are ρ_a, ρ_b and ρ_c, for which we define the quantities $g = (g_a g_b g_c)^{\frac{1}{3}}$ and $\rho = (\rho_a \rho_b \rho_c)^{\frac{1}{3}}$. The principal axis a is assumed to make an angle θ with the positive x-axis. Green's function $\mathscr{G}$ is defined by Poisson's equation

$$\nabla \cdot (\tilde{g} \nabla \mathscr{G}) = -g\delta(\mathbf{R} - \mathbf{R}_0) \tag{C.1}$$

subject to the boundary condition

$$(\tilde{g} \nabla U) \cdot \mathbf{z} = 0 \tag{C.2}$$

on the ground surface $z = 0$, and the regularity condition at infinity. The solution for $\mathscr{G}$ can be written in the following form:

$$\mathscr{G}(\mathbf{R}, \mathbf{R}_0) = \frac{1}{4\pi}\left(\frac{1}{\mathscr{R}} - \frac{1}{\mathscr{R}'}\right) \tag{C.3}$$

where $\mathbf{R}$ is the calculation point (x, y, z), and $\mathbf{R}_0$ is the source point (x_0, y_0, z_0). The functions $\mathscr{R}$ and $\mathscr{R}'$ are given as [38]

$$\mathscr{R} = [A(x-x_0)^2 + \beta(y-y_0)^2 + B(z-z_0)^2 + C(x-x_0)(y-y_0)]^{\frac{1}{2}} \tag{C.4a}$$

$$\mathscr{R}' = [A(x - x_0 - Cz_0/A)^2 + \beta(y - y_0)^2 + B(z + z_0)^2$$
$$+ C(x - x_0 - Cz_0/A)(y - y_0)]^{\frac{1}{2}} \tag{C.4b}$$

$$A = a\cos^2\theta + \gamma\sin^2\theta$$

$$B = a\sin^2\theta + \gamma\cos^2\theta$$
$$C = 2(a - \gamma)\sin\theta\cos\theta$$

$$a = g/g_a = \rho_a/\rho$$
$$\beta = g/g_b = \rho_b/\rho$$
$$\gamma = g/g_c = \rho_c/\rho$$

Appendix D

Electric Green's dyadic for a half-space below the ground surface

In this appendix we give, following Hohmann [52], the electric Green's dyadic, $\tilde{\mathbf{G}}^{2,2}$, for the ground half-space lying below an earth–air interface. The Cartesian system of coordinates is fixed such that the z-axis is directed into the earth and the earth–air interface is represented by the plane $z = 0$. Referring to Fig. 6.1, region V_2 represents the earth, V_3 the air and A the earth–air interface. Magnetic permeability in the earth is assumed to be equal to that of free space, and displacement currents are neglected, hence the wavenumber in the earth is $k^2 = i\omega g \mu_0$.

The boundary-value problem for Green's dyadic is defined in accordance with Section 6.3 by the wave equation:

$$\nabla \times \nabla \times \tilde{\mathbf{G}}(\mathbf{R}, \mathbf{R}_0) - k^2 \tilde{\mathbf{G}}(\mathbf{R}, \mathbf{R}_0) = \tilde{\mathbf{I}}\delta(\mathbf{R} - \mathbf{R}_0) \tag{D.1}$$

the boundary conditions:

$$\mathbf{z} \times \tilde{\mathbf{G}}^{2,2} = \mathbf{z} \times \tilde{\mathbf{G}}^{3,2} \tag{D.2}$$

$$\frac{1}{\mu_2} \mathbf{z} \times \nabla \times \tilde{\mathbf{G}}^{2,2} = \frac{1}{\mu_0} \mathbf{z} \times \nabla \times \tilde{\mathbf{G}}^{3,2} \tag{D.3}$$

on the plane $z = 0$, and the radiation condition

$$\nabla \times \tilde{\mathbf{G}}^{3,2} - k \frac{\mathbf{R}}{R} \times \tilde{\mathbf{G}}^{3,2} = 0\left(\frac{1}{R}\right) \tag{D.4}$$

at infinity.

Let us write the solution for equations (D.1)–(D.4) as the sum of the whole-space and the secondary dyadics:

$$\tilde{\mathbf{G}}^{2,2} = \tilde{\mathbf{G}}_0 + \tilde{\mathbf{G}}_S \tag{D.5}$$

Further, dividing the whole-space and the secondary dyadics into components generated by the solenoidal and the laminar sources, denoted by subscripts A and φ, we write:

$$\widetilde{\mathbf{G}}_0 = \mathbf{G}_{0A} + \mathbf{G}_{0\varphi} \tag{D.6}$$

$$\widetilde{\mathbf{G}}_S = \mathbf{G}_{SA} + \mathbf{G}_{S\varphi} \tag{D.7}$$

The four terms of the dyadic $\widetilde{\mathbf{G}}^{2,2}$ can now be given by the following formulas:

$$\mathbf{G}_{0A} = \frac{e^{-ik|\mathbf{R}-\mathbf{R}_0|}}{4\pi |\mathbf{R}-\mathbf{R}_0|}(\mathbf{xx}+\mathbf{yy}+\mathbf{zz}) \tag{D.8}$$

$$\mathbf{G}_{0\varphi} = \gamma_3 [(x-x_0)(\mathbf{xx}+\mathbf{xy}+\mathbf{xz})$$
$$+ (y-y_0)(\mathbf{yx}+\mathbf{yy}+\mathbf{yz})$$
$$+ (z-z_0)(\mathbf{zx}+\mathbf{zy}+\mathbf{zz})] \tag{D.9}$$

$$\mathbf{G}_{SA} = \gamma_2 \mathbf{xx} + \gamma_2 \mathbf{yy} - \frac{e^{-ik|\mathbf{R}-\mathbf{R}_0'|}}{4\pi |\mathbf{R}-\mathbf{R}_0'|} \mathbf{zz} \tag{D.10}$$

$$\mathbf{G}_{S\varphi} = [(x-x_0)(\gamma_1 \mathbf{xx}+\gamma_1 \mathbf{xy}+\gamma_4 \mathbf{xz})$$
$$+ (y-y_0)(\gamma_1 \mathbf{yx}+\gamma_1 \mathbf{yy}+\gamma_4 \mathbf{yz})$$
$$+ (z+z_0)(\gamma_4 \mathbf{zx}+\gamma_4 \mathbf{zy}+\gamma_4 \mathbf{zz})] \tag{D.11}$$

with

$$\gamma_1 = \frac{1}{4\pi r}\int_0^\infty \left(2-\frac{\lambda}{u}\right)e^{-u(z+z_0)}\lambda J_1(\lambda r)d\lambda \tag{D.12}$$

$$\gamma_2 = \frac{1}{4\pi}\int_0^\infty \left(\frac{u-\lambda}{u+\lambda}\right)\frac{\lambda}{u}e^{-u(z+z_0)}J_0(\lambda r)d\lambda \tag{D.13}$$

$$\gamma_3 = (ik|\mathbf{R}-\mathbf{R}_0|+1)\frac{e^{-ik|\mathbf{R}-\mathbf{R}_0|}}{4\pi |\mathbf{R}-\mathbf{R}_0|^3} \tag{D.14}$$

$$\gamma_4 = (ik|\mathbf{R}-\mathbf{R}_0'|+1)\frac{e^{-ik|\mathbf{R}-\mathbf{R}_0'|}}{4\pi |\mathbf{R}-\mathbf{R}_0'|^3} \tag{D.15}$$

$$u = (\lambda^2 - k^2)^{\frac{1}{2}}$$
$$r = [(x-x_0)^2 + (y-y_0)^2]^{\frac{1}{2}}$$
$$|\mathbf{R}-\mathbf{R}_0| = [r^2 + (z-z_0)^2]^{\frac{1}{2}}$$
$$|\mathbf{R}-\mathbf{R}_0'| = [r^2 + (z+z_0)^2]^{\frac{1}{2}}$$

References

1. Eloranta, E. H. (1986) *Acta Polytechnica Scandinavica*, Applied Physics Series, no. 153.
2. Colton, D. and Kress, R. (1983) *Integral Equation Methods in Scattering Theory*, John Wiley & Sons, New York.
3. Mikhlin, S. G. and Prossdorf, S. (1986) *Singular Integral Operators*, Springer-Verlag, Berlin.
4. Petrovsky, I. G. (1975) *Lectures on the Theory of Integral Equations*, Mir Publishers, Moscow.
5. Harrington, R. G. (1968) *Field Computation by Moment Methods*, Macmillan, New York.
6. Stratton, J. A. (1941) *Electromagnetic Theory*, McGraw-Hill, New York.
7. Roach, G. F. (1982) *Green's Functions*, Cambridge University Press, Cambridge.
8. Kellogg, O. D. (1967) *Foundations of Potential Theory*, Springer-Verlag, Berlin.
9. Parasnis, D. S. (1986) *Principles of Applied Geophysics*, Chapman and Hall, London.
10. Ryss, U. S. (1981) *Geoexploration*, **18**, 281–295.
11. Schulz, R. (1985) *J. Geophys*, **56**, 192–200.
12. Soininen, H. (1985) *Geophysical Transactions*, **31**, (4), 359–371.
13. Eloranta, E. H. (1986) *Geophys. Prosp.*, **34**, 856–872.
14. Snyder, D. D. (1976) *Geophysics*, **41**, 997–1015.
15. Eskola, L. and Hongisto, H. (1981) *Geophys. Prosp.*, **29**, 260–273.
16. Eskola, L. (1984) *Geophys. Prosp.*, **32**, 510–511.
17. Hongisto, H. (1986) The accuracy of the numerical solution used in a method for resistivity and IP modelling of $2\frac{1}{2}$-dimensional bodies (in Finnish), unpublished thesis, Helsinki University of Technology.
18. Eskola, L., Soininen, H. and Oksama, M. (1989) *Geoexploration*, **26**, 95–104.
19. Edwards, R. N., Lee, H. and Nabighian, M.N. (1978) *Geophysics*, **43**, 1176–1203.
20. Parasnis, D. S. (1988) *Geoexploration*, **25**, 177–198.
21. Soininen, H. (1987) *Geoexploration*, **24**, 455–460.
22. Eloranta, E. H. (1986) *Geoexploration*, **24**, 1–14.
23. Guptasarma, D. (1983) *Geophysics*, **48**, 98–106.

24. Eskola, L., Eloranta, E. and Puranen, R. (1984) *Geophys. Prosp.*, **32**, 79–87.
25. Eskola, L. (1986) *Geophysics*, **50**, 1638–1639.
26. Sumner, J. S. (1976) *Principles of Induced Polarization for Geophysical Exploration*, Elsevier, Amsterdam.
27. Fuller, B. D. and Ward, S. H. (1970) *IEEE Transactions on Geoscience Electronics*, **GE-8**, 7–18.
28. Papoulis, A. (1962) *The Fourier Integral and its Applications*, McGraw-Hill, New York.
29. Seigel. H. O. (1974) *Geophysics*, **39**, 321–339.
30. Pelton, W. H., Ward, S. H., Hallof, P.G., Sill, W. R. and Nelson, P. H. (1978) *Geophysics*, **43**, 588–609.
31. Soininen, H. (1984) *Geophysics*, **49**, 1534–1540.
32. Soininen, H. (1985) *Geophysics*, **50**, 810–819.
33. Telford, W. M., Geldart, L. P., Sheriff, R. E. and Keys, D. A. (1976) *Applied Geophysics*, Cambridge University Press, Cambridge.
34. Sato, M. and Mooney, H. M. (1960) *Geophysics*, **25**, 226–249.
35. Eskola, L. and Hongisto, H. (1987) *Geoexploration*, **24**, 219–226.
36. Fitterman, D. V. (1979), *Geophysics*, **44**, 195–205.
37. Eskola, L. (1988) *Geoexploration*, **25**, 211–217.
38. Eloranta, E. H. (1988) *Geoexploration*, **25**, 93–101.
39. Kobayashi, M. (1978) *IEEE Transactions on Microwave Theory and Techniques*, **MTT-26** (7), 510–512.
40. Tabarovskii, L. A. (1977) *Geologiya i Geofizika*, **18** (5), 81–88.
41. Eskola, L. and Tervo, T. (1980) *Geoexploration*, **18**, 79–95.
42. Eskola, L. (1984) *Geoexploration*, **22**, 75.
43. Tervo, T. (1983) On the application of the method of sub- areas in calculating magnetostatic effects (in Finnish), unpublished thesis, University of Oulu.
44. Doherty, J. (1988) *Geophys. Prosp.*, **36**, 644–668.
45. Muller, C. (1969) *Foundations of the Mathematical Theory of Electromagnetic Waves*, Springer-Verlag, New York.
46. Tai, C. -T. (1987) *Radio Science*, **22** (7), 1283–1288.
47. Tai, C. -T. (1971) *Dyadic Green's Functions in Electromagnetic Theory*, International Textbook Co., London.
48. Levine, H. and Schwinger, J. (1950) *Comm. Pure and Appl. Math.*, **III**, 355–391.
49. Weidelt, P. (1975) *J. Geophys.*, **41**, 85–109.
50. Wannamaker, P. E., Hohmann, G. W. and SanFilipo, W. A. (1984) *Geophysics*, **49**, 60–74.
51. Raiche, A. and Coggon, J. (1975) *Geophys. J. R. Astr. Soc.*, **42**, 1035–1038.
52. Hohmann, G. W. (1975) *Geophysics*, **40**, 309–321.
53. Oksama, M. (1980) Calculation of electromagnetic anomalies by means of the integral equation method (in Finnish), unpublished thesis, Helsinki University of Technology.
54. Tai, C. -T. (1985), *Electromagnetics*, **5**, 79–88.
55. Raiche, A. P. (1974) *Geophys. J. R. Astr. Soc.*, **36**, 363–376.
56. Van Bladel, J. (1961) *IRE Trans. Antennas Propagation*, **9**, 563–566.
57. Ting, S. C. and Hohmann, G. W. (1981) *Geophysics*, **46**, 182–197.
58. Yaghjian, A. (1980) *Proceedings of the IEE*, **68** (2), 248–263.
59. Hohmann, G. W. (1988) Numerical modeling for electromagnetic methods of

geophysics, in *Electromagnetic methods in applied geophysics-theory*, Vol. 1, ed. by M. N. Nabighian, Society of Exploration Geophysicists, Tulsa, Oklahoma.
60. Newman, G. A. and Hohmann, G. W. (1988) *Geophysics*, **53**, 691–706.
61. Lee, K. H. and Morrison, H. F. (1985) *Geophysics*, **50**, 1163–1165.
62. Lee, K. H. and Morrison, H. F. (1984) A solution for TM-mode plane waves incident on a two-dimensional inhomogeneity, Lawrence Berkeley Laboratory, Report LBL-17857.
63. Hohmann, G. W. (1971) *Geophysics*, **36**, 101–131.
64. Parry, J. R. and Ward, S. H. (1971) *Geophysics*, **36**, 67–100.
65. Lajoie, J. J. and West, G. F. (1976) *Geophysics*, **41**, 1133–1156.
66. Dyck, A. V., Bloore, M. and Vallee, M. A. (1981), User manual for programs PLATE and SPHERE, Research in Applied Geophysics, 14, Geophysics Laboratory, Dept of Physics, University of Toronto.
67. Morgan, T. R. (1983) *Foundations of Wave Theory for Seismic Exploration*, International Human Resources Development Corporation, Boston.
68. Pao, Y. H. and Varatharajulu, V. (1976) *J. Acoust. Soc. Am.*, **59**, 1361–1371.
69. Berkhout, A. J. (1987) *Applied Seismic Wave Theory*, Elsevier, Amsterdam.
70. Wapenaar, C. P. A. and Haimé, G. C. (1990) *Geophys. Prosp.*, **38**, 23–60.
71. Kuhn, M. J. and Alhilali, K. A. (1977), *Geophysics*, **42**, 1183–1198.
72. Heikkinen, P. (1984), Seismic reflection sounding on a bedrock area: results from an experiment in Ylivieska (in Finnish), unpublished thesis, University of Helsinki.
73. Trorey, A. W. (1970) *Geophysics*, **35**, 762–784.
74. French, W. S. (1975) *Geophysics*, **40**, 961–980.
75. Van Nostrand, R. G. and Cook, K. L. (1966) *US Geological Survey, Professional Papers*, **499**, 134–135.

Index

Acoustic waves, *see* Compressional wave field
Anisotropy
  elastic 159–63
  electric 87–93
Apparent resistivity 23, 135

Basis functions 6
Biot and Savart's law 49
Boundary conditions for
  compressional wave fields
    Dirichlet or free-surface 173
    Neumann or rigid-surface 173–4
  electromagnetic fields 10
  Green's dyadics
    electromagnetic 129
  Green's functions
    electric 13, 88
    magnetic 96, 121
  scalar potentials
    electric 12, 88
    magnetic 96

Charge
  conservation of 34, 45–6, 54
  line 1–2
  surface 17–18, 90
  volume 16, 85
Cole–Cole dispersion model 70
Compressional wave field 169–70
  migration of 175–6
  scattering of 174–5

  velocity of 165
Conductance 44, 154
Conductivity
  anisotropy 82–3
  of rocks 22–3
Constitutive relations
  for elastic fields 160, 163
  for electromagnetic fields 83, 94, 121, 124
  see also Ohm's law
Contact polarization method 23
Convergence of numerical solution 27, 41, 113
Curie point 102
Current
  channeling 46–8
  conservation of 4, 62, 88, 90
  electrode
    line 1–4
    point 15–16, 84–5

Demagnetization 110–13
  factor 112
  tensor 111
Depolarization dyadic 133–4
Dispersion of resistivity 22, 66–8
Displacement tensor 160, 164
Dyadic Green's function, *see* Green's dyadic

Elastic
  constants 160, 163

waves 164, 168–70
Electric
  anisotropy
    definition 82–3
    physical significance of 83–7
    principal axes of 83
  conductivity, see Conductivity
  double layer 18–20, 60–61, 74–5, 79
  methods 21–2
  polarization 22
Electrode polarization 22
Electrostatic saturation 33
Equivalent sources 94, 127, 145

Fictitious sources 32, 56–7, 81–2, 90, 91
Fourier transform 37, 39, 66–7, 148
Fredholm integral equation 5

Galerkin method 7
Gauss's
  law 10, 95
  theorem 161
Green's
  dyadic 128–30
    for half-space 143, 184–5
    symmetry of 132
    for whole-space 130
  function 5, 13
    for anisotropic half-space 182–3
    for anisotropic whole-space 90
    of a double source 19
    for half-space 2, 104, 107, 173, 174
    for half-space with a vertical contact 180–81
    reciprocity of 32, 89, 170
    for two-layer half-space 177–9
    for whole-space 13, 96, 107, 165, 173
  second identity 13
    for anisotropic medium 88–9
  vector-dyadic 128
  tensors 160, 163–4
    symmetry of 160

Helmholtz integral formulas
  for acoustic fields 171
  for full elastic fields 161, 166
High-susceptibility models 103–10

Hilbert transform 67–8

Induced polarization
  information-theoretical relations 66–8
  modelling 69–70
Interfacial impedance, see Surface impedance

Kernel 5
  see also Green's function
Kirchhoff integral formula 172–3

Lamé
  constants 163
  potentials 168
Laminar field 10, 154
  see also Compressional wave field
Longitudinal conductance 44, 154
Low-susceptibility models 113–16

Magnetic
  dipoles 100–1
  IP (MIP) 70
  moment of a body 112
  poles
    surface 99–100
    volume 97–8
  properties of rocks 102–3
  susceptibility 102
Magnetization
  induced 100
  remanent 120
  see also Remanence
Magnetometric resistivity 48–51
Maxwell's equations
  time domain 9–10
  frequency domain 124
Membrane polarization 22
Migration 175–6
Mise-à-la-masse method 52–9
MMR, see Magnetometric resistivity
Mode separation of
  elastic fields 168
  electromagnetic fields 139–40, 155
Moment method 6–7

Numerical methods 6–8

Ohm's law 66, 67, 69, 83

Perfect conductor
  in anisotropic environment 87–91
  in isotropic environment 33–5, 45–6, 57–9, 61–2, 75–6
Point-matching method 7–8
Principal axes
  of conductivity 83
  of demagnetization tensor 112
  of resistivity 83
Propagation constant 125
P-waves, *see* Compressional wave field

Radiation conditions
  for compressional wave fields 170
  for full elastic wave fields 166
  for electromagnetic wave fields 125, 145
  for Green's dyadics 129
Rayleigh–Sommerfeld integral formulas 173–4
Reciprocal electrode configuration 53–4
Regularity condition for potential 12, 88
Relaxation methods 21
Remanence 120–2
Resistivity
  anisotropy 83
  dispersion 22, 66–8
  methods 23–48
  nonlinearity of 22–3
  of rocks 22–3
  tensor 83
Rotational wave field 168–9

Seismic
  methods 158–9
  velocities 165

Self-potentials
  macroscopic models
    for electrofiltration potential 78–82
    for mineralization potential 73–8
  primary potentials 75, 79
  scaling of 75–6
  origin of 71–3
Shear (S) wave field, *see* Rotational wave field
Solenoidal field 155, 168
Spectral IP 69–71
  see also Resistivity dispersion
Streaming potential 71–2
Stress tensor 160, 164
Subsection method 7–8, 27
Surface
  charge 17–18, 90
  impedance 18, 61, 63
  polarization 59–60
  polarization model 60–64
  scaling of 64–6

Test functions 7
Thin sheet model
  electric 42–8
  electromagnetic 153–7
  magnetic 107–10, 114–16
$2\frac{1}{2}$-dimensional models
  electric 36–42
  electromagnetic 148–53
  magnetic 106–7
  sampling of spatial wavenumbers 38–9, 40, 153
Two-dimensional models 139–44, 165

Volume polarization 60
  see also Resistivity dispersion